Electrical installation calculations

VOLUME 1

A J Watkins
B.Sc., B.Sc.(Aston), C.Eng., MIEE

SIXTH EDITION
prepared by

R. K. Parton
Formerly the Head of Department of Electrical and Motor Vehicle Engineering,
The Reid Kerr College, Paisley

BUTTERWORTH
HEINEMANN

OXFORD   AUCKLAND   BOSTON   JOHANNESBURG   MELBOURNE   NEW DELHI

Butterworth-Heinemann
Linacre House, Jordan Hill, Oxford OX2 8DP
225 Wildwood Avenue, Woburn, MA 01801-2041
A division of Reed Educational and Professional Publishing Ltd

R A member of the Reed Elsevier plc group

First published by Arnold in 1957
Sixth edition by Arnold 1988
Reprinted by Butterworth-Heinemann 2001

British Library Cataloguing in Publication Data
A catalogue record for this book is available from the British Library

ISBN 0 340 73184 2

Printed by St Edmundsbury Press Ltd, Bury St Edmunds, Suffolk

Contents

Preface to the sixth edition

This book is the first of a series of three volumes intended for students of electrical installation work. They are essentially books of examples aimed to co-ordinate the technology and calculations needed to achieve an NVQ in electrical installation competences and to support recent syllabuses of City and Guilds Courses 236, 2380, 2391, 2400 and 2351 and also similar electrical installation courses of other examination bodies such as BTEC and SCOTVEC. Representative examples are worked and a selection of problems, with answers, is provided to help the student to practise the techniques involved.

A small number of typical multiple-choice questions are included: to answer these the student should work through the question in the usual way and select which of the given answers corresponds to the result they obtained. It should be noted that each of the incorrect answers given is the result of using a wrong technique or a miscalculation. A student selecting the wrong answer should therefore check that he or she is using the correct method or making the correct calculation.

In this the sixth edition, the content has been revised to take account of the adoption by the British Standards Institution of the Institution of Electrical Engineers Wiring Regulations, Sixteenth edition to form BS 7671: 1992.

This book takes into consideration Amendments No. 1, 1994 and No. 2, 1997 to British Standard BS 7671. Readers are advised to have as additional reference material one of the site-type electricians' guides, e.g. the IEE On-Site Guide.

Throughout the book the 'preferred' symbol U is used to signify potential difference (p.d), thus U indicates the quantity of supply voltage, i.e. 230 V (i.e. 230 units of voltage).

An up-date of electric tariff charges, materials costs and VAT levels has been undertaken, but it is obvious that such costs escalate rapidly and the printed figures should be recognized as typical values at the time of revision.

The publishers are grateful to the British Standards Institution and the Institution of Electrical Engineers for permission to make use of data from their publications. Gratitude is also expressed to the City and Guilds for permission to use various questions from past examinations, in certain cases slight re-wording was necessary to satisfy modern terminology and requirements. The two institutions and the City and Guilds however accepts no responsibility for the interpretation by the author of their specific requirements or indeed for the answers to the various questions. The City and Guilds accepts no responsibility for any of the answers quoted for these questions.

The revision has retained the 'pocket size' format whilst introducing more white space to reduce the previous cramped feeling.

R.K.P.
Kilmacolm
1998

Solving typical simple equations

Type 1: Ohm's law, resistance of conductors, electric power, etc.

(a) $3x = 12$ (b) $\dfrac{x}{3} = 12$ (c) $\dfrac{3}{x} = 12$

$\therefore \quad x = \dfrac{12}{3}$ $\therefore \quad x = 3 \times 12$ $\therefore \quad \dfrac{x}{3} = \dfrac{1}{12}$

$\therefore \quad x = 4$ $\therefore \quad x = 36$ $\therefore \quad x = \dfrac{3}{12} = \dfrac{1}{4}$

$= 0.25$

Type 2: resistors in series, heat calculations, etc.

(a) $x + 3 = 12$ (b) $x - 3 = 12$

$\therefore \quad x = 12 - 3$ $\therefore \quad x = 12 + 3$

$\therefore \quad x = 9$ $\therefore \quad x = 15$

Type 3: resistors in parallel, etc.

$$3 + \frac{3}{x} = 12$$

$\therefore \quad \dfrac{3}{x} = 12 - 3$

$\therefore \quad \dfrac{3}{x} = 9$

$\therefore \quad \dfrac{x}{3} = \dfrac{1}{9}$

$\therefore \quad x = \dfrac{3}{9} = \dfrac{1}{3}$

$= 0.33$

Type 4: heat calculations, etc.

(a) $2x - 3 = 12$

$\therefore \qquad 2x = 12 + 3$

$\therefore \qquad 2x = 15$

$\therefore \qquad x = \dfrac{15}{2}$

$\qquad\qquad = 7.5$

(b) $2(x - 3) = 12$

$\therefore \qquad (x - 3) = \dfrac{12}{2}$

$\therefore \qquad x - 3 = 6$

$\therefore \qquad x = 6 + 3$

$\qquad\qquad = 9$

Multiples and submultiples of SI units

mega (M) unit multiplied by 1 000 000

e.g. 1 megawatt = 1 000 000 watts

$$1\,\text{MW} = 10^6\,\text{W}$$

kilo (k) unit multiplied by 1000

e.g. 1 kilogramme = 1000 grammes

$$1\,\text{kg} = 10^3\,\text{g}$$

deci (d) unit divided by 10

e.g. 1 metre = 10 decimetres

or $1\,\text{dm} = 10^{-1}\,\text{m}$

centi (c) unit divided by 100

e.g. 1 litre = 100 centilitres

or $1\,\text{cl} = 10^{-2}\,\text{l}$

milli (m) unit divided by 1000

e.g.   1 ampere = 1000 milliamperes

$$1\,A = 1000\,mA$$

or $1\,mA = 10^{-3}\,A$

micro (μ) unit divided by 1000 000

e.g.   1 coulomb = 1000 000 microcoulombs

$$1\,C = 10^6\,\mu C$$

or $1\,\mu C = 10^{-6}\,C$

nano (n) unit divided by 1000 000 000

e.g.   1 second = 1000 000 000 n seconds

or $1\,ns = 10^{-9}\,s$

pico (p) unit divided by 1000 000 000 000

e.g.   1 farad = 1000 000 000 000 picofarad

or $1\,pF = 10^{-12}\,F$

EXAMPLE 1   Convert 1.125 kilowatts to watts

$$1.125\,kW = 1.125\,\cancel{kW}\left[\frac{1000\,W}{1\,\cancel{kW}}\right]$$

$$= 1.125 \times 1000\,W$$

$$= 1125\,W$$

(Note use of unity bracket and cancelling of units.)

EXAMPLE 2   Convert 25 microcoulombs to coulombs.

$$25\,\mu C = 25\,\cancel{\mu C}\left[\frac{1\,C}{10^6\,\cancel{\mu C}}\right]$$

$$= \frac{25}{10^6}\,C$$

$$= 25 \times 10^{-6}\,C \quad \text{or} \quad 0.000\,025\,C$$

EXAMPLE 3 Convert 0.0006 microfarad to picofarads.

$$1 \text{ farad (F)} = 10^6 \text{ microfarads } (\mu F)$$

$$= 10^{12} \text{ picofarads (pF)}$$

thus $\qquad 10^6 \, \mu F = 10^{12} \, pF$

$$1 \, \mu F = \frac{10^{12}}{10^6} \, pF$$

$$= 10^6 \, pF$$

$$0.0006 \, \mu F = 0.0006 \, \cancel{\mu F} \left[\frac{10^6 \, pF}{1 \, \cancel{\mu F}} \right]$$

$$= 0.0006 \times 10^6 \, pF$$

$$= 600 \, pF$$

EXERCISE 1

1. Convert 2.768 kW to watts.
2. How many ohms are there in 0.45 MΩ?
3. Express a current of 0.037 A in milliamperes.
4. Convert 3.3 kV to volts.
5. Change 0.000 596 MΩ to ohms.
6. Find the number of kilowatts in 49 378 W.
7. The current in a circuit is 16.5 mA. Change this to amperes.
8. Sections of the 'Grid' system operate at 132 000 V. How many kilovolts is this?
9. Convert 1.68 μC to coulombs.
10. Change 724 mW to watts.
11. Convert the following resistance values to ohms:
    (a) 3.6 μΩ   (d) 20.6 μΩ
    (b) 0.0016 MΩ   (e) 0.68 μΩ
    (c) 0.085 MΩ

12. Change the following quantities of power to watts:
    (a) 1.85 kW
    (b) 18.5 mW
    (c) 0.185 MW
    (d) 1850 μW
    (e) 0.0185 kW

13. Convert to volts:
    (a) 67.4 mV
    (b) 11 kV
    (c) 0.240 kV
    (d) 9250 μV
    (e) 6.6 kV

14. Convert the following current values to amperes:
    (a) 345 mA
    (b) 85.4 μA
    (c) 29 mA
    (d) 0.5 mA
    (e) 6.4 mA

15. Add the following resistances together and give the answer in ohms:
    18.4 Ω, 0.000 12 MΩ, 956 000 μΩ

16. The following items of equipment are in use at the same time: four 60 W lamps, two 150 W lamps, a 3 kW immersion heater, and a 1.5 kW radiator. Add them to find total load and give the answer in watts.

17. Express the following values in more convenient units:
    (a) 0.0053 A
    (b) 18 952 W
    (c) 19 500 000 Ω
    (d) 0.000 006 25 C
    (e) 264 000 V

18. The following loads are in use at the same time: a 1.2 kW radiator, 15 W lamp, a 750 W iron, and a 3.5 kW washing machine. Add them together and give the answer in kilowatts.

19. Add 34 250 Ω to 0.56 MΩ and express the answer in ohms.

20. From 25.6 mA tale 4300 μA and give the answer in amperes.

21. Convert 32.5 μC to coulombs.

22. Convert 4350 pF to microfarads.

23. 45 μs is equivalent to:
    (a) 0.45 s   (b) 0.045 s   (c) 0.0045 s   (d) 0.000 045 s

24. 50 cl is equivalent to:
    (a) 5 l   (b) 0.05 l   (c) 0.05 ml   (d) 500 ml

25. 0.2 m³ is equivalent to:
    (a) 200 dm³   (b) 2000 cm³   (c) 2000 dm³   (d) 200 cm³

26. 0.6 MΩ is equivalent to:
    (a) 6000 Ω   (b) 60 000 Ω   (c) 600 000 Ω   (d) 6000 000 Ω

5

Circuit calculations

Note This book employs the symbol '*U*' for voltage quantity (*potential difference*) and the symbol '*V*' for unit voltage. However the symbol '*V*' is still widely used to represent both voltage quantity and the unit of voltage.

OHM'S LAW

In a d.c. circuit (Fig. 1), the current is directly proportional to the applied voltage and inversely proportional to the resistance:

$$\text{voltage} = \text{current} \times \text{resistance}$$

$$U = I \times R$$

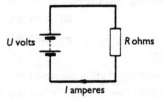

Fig. 1

EXAMPLE 1   The current in a circuit is 100 A and the resistance is 0.1 Ω. Find the voltage.

$$U = I \times R$$
$$= 100 \times 0.1$$
$$= 10\,\text{V}$$

EXAMPLE 2   The voltage applied to a circuit is 100 V and the current flowing is 15 A. Calculate the resistance.

$$U = I \times R$$
$$\therefore \quad 100 = 15 \times R$$
$$\therefore \quad \frac{100}{15} = R$$
$$\therefore \quad R = 6.666$$
$$= 6.67\,\Omega \quad \text{(correct to three significant figures)}$$

EXAMPLE 3 A voltage of 230 V is applied to a resistor of 25 Ω. Calculate the current which flows.

$$230 = I \times R$$
$$230 = I \times 25$$
$$\therefore \quad \frac{230}{25} = I$$
$$\therefore \quad I = 9.2\,\text{A}$$

In a.c. circuits (Fig. 2), the current is limited by the impedance (Z). Impedance is measured in ohms, and

voltage = current (amperes) × impedance (ohms)
$$U = I \times Z$$

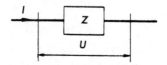

Fig. 2

EXAMPLE 4 The current through an impedance of 36 Ω is 7 A. Calculate the voltage drop.

$$U = I \times Z$$
$$= 7 \times 36$$
$$= 252\,\text{V}$$

EXAMPLE 5 A current of 8 A flows in an a.c. magnet circuit connected to a 230 V supply. Determine the impedance of the magnet coil.

$$U = I \times Z$$

$$\therefore \quad 230 = 8 \times Z$$

$$\therefore \quad Z = \frac{230}{8}$$

$$= 28.75\,\Omega$$

RESISTORS IN SERIES

When a number of resistors are connected *in series* (Fig. 3), the total resistance is equal to *the sum* of the resistance values.

If R is the total resistance, then

$$R = R_1 + R_2 + R_3 + \ldots + \text{etc.}$$

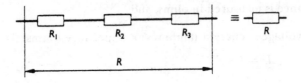

Fig. 3

EXAMPLE 1 Resistors of $0.413\,\Omega$, $1.275\,\Omega$, and $0.896\,\Omega$ are connected in series. Find the total resistance.

$$R = R_1 + R_2 + R_3$$

$$= 0.413 + 1.275 + 0.896$$

$$= 2.584\,\Omega$$

EXAMPLE 2 Find the value of a resistor which, when connected in series with one of $4.6\,\Omega$, will make a total of $5.5\,\Omega$.

$$R = R_1 + R_2$$
$$5.5 = 4.6 + R_2$$
$$\therefore \quad R_2 = 5.5 - 4.6$$
$$= 0.9\,\Omega$$

EXERCISE 2

1. Calculate the total resistance of each of the following groups of resistors in series. (*Values are in ohms unless otherwise stated.*)

   (a) 12, 35, 59
   (b) 8.4, 3.5, 0.6
   (c) 19.65, 4.35
   (d) 0.085, 1.12, 0.76
   (e) 27.94, 18.7, 108.3
   (f) 256.5, 89.7
   (g) 1400, 57.9 kΩ
   (h) 1.5 MΩ, 790 000
   (i) 0.0047, 0.095
   (j) 0.0568, 0.000 625 (*give answers in microhms*)

2. Determine the value of resistance which, when connected in series with the resistance given, will produce the required total.

   (a) 92 Ω to produce 114 Ω
   (b) 12.65 Ω to produce 15 Ω
   (c) 1.5 Ω to produce 3.25 Ω
   (d) 4.89 Ω to produce 7.6 Ω
   (e) 0.9 Ω to produce 2.56 Ω
   (f) 7.58 Ω to produce 21 Ω
   (g) 3.47 Ω to produce 10 Ω
   (h) 195 Ω to produce 2000 Ω
   (i) 365 $\mu\Omega$ to produce 0.5 Ω (*answer in microhms*)
   (j) 189 000 Ω to produce 0.25 MΩ (*answer in megohms*)

3. Calculate the total resistance when four resistors each of 0.84 Ω are wired in series.

4. Resistors of 19.5 Ω and 23.7 Ω are connected in series. Calculate the value of a third resistor which will give a total of 64.3 Ω.

5. How many 0.58 Ω resistors must be connected in series to make a total resistance of 5.22 Ω?

6. A certain type of lamp has a resistance of 41 Ω. What is the resistance of 13 such lamps in series? How many of these lamps are necessary to make a total resistance of 779 Ω?

7. The four field coils of a motor are connected in series and each has a resistance of 33.4 Ω. Calculate the total resistance. Determine also the value of an additional series resistance which will give a total resistance of 164 Ω.

8. Two resistors connected in series have a combined resistance of 4.65 Ω. The resistance of one of them is 1.89 Ω. What is the resistance of the other?

9. Four equal resistors are connected in series and their combined resistance is 18.8 Ω. The value of each resistor is

   (a) 9.4 Ω   (b) 75.2 Ω   (c) 4.7 Ω   (d) 37.6 Ω

10. Two resistors connected in series have a combined resistance of 159 Ω. One resistor has a value of 84 Ω. The value of the other is

    (a) 133.56 Ω   (b) 1.89 Ω   (c) 243 Ω   (d) 75 Ω

11. Two resistors of equal value are connected to three other resistors of value 33 Ω, 47 Ω and 52 Ω to form a series group of resistors with a combined resistance of 160 Ω.

    What is the resistance of the two unknown resistors?

    (a) 7 Ω   (b) 14 Ω   (c) 28 Ω   (d) 42 Ω

12. Four resistors of value 23 Ω, 27 Ω, 33 Ω, 44 Ω are connected in series. It is required to modify their combined resistance to 140 Ω by replacing one of the existing resistors by a new resistor of value 40 Ω. Which of the original resistors should be replaced?

    (a) 23 Ω   (b) 27 Ω   (c) 33 Ω   (d) 44 Ω

CONDUCTOR RESISTANCE

Any conductor used to carry current in an electrical circuit will possess a certain resistance to the current flow. The value of this resistance is determined by the type of material it is made from, its cross-sectional area and its length. For a given conductor its resistance will increase directly proportional to its length.

CIRCUIT CONDUCTOR RESISTANCE AND VOLTAGE DROP

The higher the resistance of a circuit conductor, the greater the voltage drop will be along its length. The actual voltage drop in a particular circuit's conductors could be determined by the use of Ohm's law using the circuit current in amperes, the conductor's resistance in milliohms per metre and the conductor's total length. For practical purposes, however, we may employ a formula based upon the length of cable run in metres, the current carried by the cable and the type of cable conductor material. Prepared tables list the voltage drop in terms of millivolts per ampere per metre of circuit length (mV/A/m). The voltage drop listed is for conductor feed and return, e.g. in the case of single-phase a.c. or two-wire d.c. circuits, for two single-core cables or for one two-core cable.

EXAMPLE A low-voltage radial circuit is arranged as shown in Fig. 4 and is wired throughout with 70 mm^2 copper cable. If for this size of cable a voltage drop of 0.69 mV per ampere per metre occurs, calculate

(a) the current in each section,

(b) the voltage drop in each section,

(c) the supply voltage U_s.

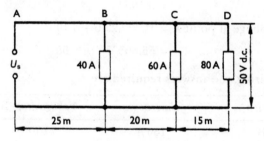

Fig. 4

For the section CD,

$$\text{current} = 80\,\text{A}$$

$$\text{voltage drop} = 0.69\,\frac{\text{mV}}{\text{A\,m}} \times 80\,\text{A} \times 15\,\text{m}$$

$$= 828\,\text{mV}$$

$$= 0.828\,\text{V}$$

$$\text{voltage at point C} = 50 + 0.828$$

$$= 50.828\,\text{V}$$

For the section BC,

$$\text{current} = 80 + 60$$
$$= 140\,\text{A}$$
$$\text{voltage drop} = 0.69\,\frac{\text{mV}}{\text{A\,m}} \times 140\,\text{A} \times 20\,\text{m}$$
$$= 1932\,\text{mV}$$
$$= 1.932\,\text{V}$$
$$\text{voltage at point B} = 50.828 + 1.932$$
$$= 52.76\,\text{V}$$

For the section AB,

$$\text{current} = 80 + 60 + 40$$
$$= 180\,\text{A}$$
$$\text{voltage drop} = 0.69\,\frac{\text{mV}}{\text{A\,m}} \times 180\,\text{A} \times 25\,\text{m}$$
$$= 3105\,\text{mV}$$
$$= 3.105\,\text{V}$$
$$\text{voltage at point A} = U_s = 52.76 + 3.105$$
$$= 55.865\,\text{V} \quad \text{or} \quad 55.9\,\text{V}$$

Summarizing, the answers required are

Section	Current (a)	Voltage drop (b)
AB	180 A	3.105 V
BC	140 A	1.932 V
CD	80 A	0.828 V

and the supply voltage is 55.9 V (c).

RESISTORS IN PARALLEL

When several resistors are connected *in parallel* (Fig. 5), the equivalent resistance R is given by

$$\frac{1}{R} = \frac{1}{R_1} + \frac{1}{R_2} + \frac{1}{R_3} + \ldots + \text{etc.}$$

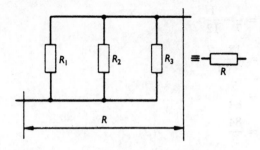

Fig. 5

Resistors of $2\,\Omega$, $4\,\Omega$, and $6\,\Omega$ are connected in parallel. Calculate the equivalent resistance.

$$\frac{1}{R} = \frac{1}{R_1} + \frac{1}{R_2} + \frac{1}{R_3}$$

$$= \frac{1}{2} + \frac{1}{4} + \frac{1}{6}$$

$$= \frac{6+3+2}{12}$$

$$= \frac{11}{12}$$

$$\therefore \quad R = \frac{12}{11}$$

$$= 1.09\,\Omega$$

EXAMPLE 2 Calculate the value of a resistor which, when connected in parallel with a $12\,\Omega$ resistor, will give a combined resistance of $7\,\Omega$.

$$\frac{1}{R} = \frac{1}{R_1} + \frac{1}{R_2}$$

$$\therefore \quad \frac{1}{7} = \frac{1}{12} + \frac{1}{R_2}$$

$$\therefore \quad \frac{1}{R_2} = \frac{1}{7} - \frac{1}{12}$$

$$= \frac{12 - 7}{84}$$

$$= \frac{5}{84}$$

$$\therefore \quad R_2 = \frac{84}{5}$$

$$= 16.8\,\Omega$$

EXAMPLE 3   A resistor of $36\,\Omega$ and one of $4\,\Omega$ are connected in parallel and between them they carry a total current of $5\,\text{A}$ (Fig. 6). Calculate (a) the voltage drop across the resistors, (b) the actual current flowing in each resistor.

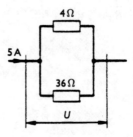

Fig. 6

(a) Let the resistance equivalent to the two in parallel be R; then

$$\frac{1}{R} = \frac{1}{R_1} + \frac{1}{R_2}$$

$$= \frac{1}{36} + \frac{1}{4}$$

$$= \frac{1 + 9}{36}$$

$$= \frac{10}{36}$$

$$\therefore \quad R = \frac{36}{10}$$

$$= 3.6\,\Omega$$

The equivalent circuit may now be drawn (Fig. 7):

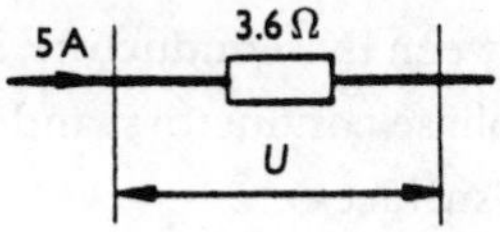

Fig. 7

Voltage drop $U = I \times R$

$$= 5 \times 3.6$$

$$= 18\,V$$

(b) To find the current flowing in each resistor, use the equation

$$U = I \times R$$

When $R = 36\,\Omega$,

$$18 = I \times 36$$

$\therefore \qquad I = \dfrac{18}{36} = 0.5\,A$

When $R = 4\,\Omega$,

$$18 = I \times 4$$

$\therefore \qquad I = \dfrac{18}{4} = 4.5\,A$

Adding as a check,

$$I = 0.5 + 4.5 = 5\,A$$

Summarizing, current through $36\,\Omega$ resistor $= 0.5\,A$
and current through Ω resistor $= 4.5\,A$

CABLE CONDUCTOR AND INSULATION RESISTANCE

In an earlier section of this book it was said that the resistance of a conductor increased directly proportional to its length, i.e. 100 m of conductor will have twice the resistance of 50 m of the same conductor.

If this conductor is covered with insulation then the insulation may allow a small amount of current to leak from the conductor to earthed metalwork, e.g. metal conduit or in the case of a mineral insulated cable from the conductors to the metal sheath. Leakage will also take place between live conductors, i.e. phase and neutral conductors, between phase conductors, and also between phase conductors and damp surfaces etc.

Insulation resistance may be likened to a group of multiple parallel resistors all capable of carrying leakage current under certain conditions. This leakage may take place along the cable's entire length, or only where the cable is exposed to moisture, or where the cable insulation is removed for termination. Thus insulation resistance may be considered to consist of a group of parallel resistors. It is impossible to predict the final insulation resistance of a cable or indeed of a circuit or of an entire installation. However, it is safe to assume that the longer the cable or the larger the installation the lower the anticipated insulation resistance will be.

EXAMPLE   A 200 m reel of single-core mineral insulated cable is tested and its conductor resistance is found to be $1.6\,\Omega$ and its insulation resistance to be $20\,\mathrm{M\Omega}$. What will be the actual conductor resistance and the anticipated insulation resistance values of 75 m of this cable?

$$\text{Actual conductor resistance} = \frac{75}{200} \times 1.6 = 0.6\,\Omega$$

$$\text{anticipated insulation resistance} = \frac{200}{75} \times 20 = 53.33\,\mathrm{M\Omega}$$

EXERCISE 3

1. The following groups of resistors are connected in parallel. In each case calculate the equivalent resistance. Where necessary, make the answers correct to three significant figures. (*All values are in ohms.*)

   (a) 2, 3, 6   (f) 14, 70
   (b) 3, 10, 5   (g) 12, 12
   (c) 9, 7   (h) 15, 15, 15
   (d) 4, 6, 9   (i) 40, 40, 40, 40
   (e) 7, 5, 10

2. In each case, calculate the value of a resistor which, when connected in parallel with the given resistor, will produce the value asked for. (*Give answers correct to three significant figures.*)

	Given resistance Ω	Resistance required Ω
(a)	48	12
(b)	20	5
(c)	9	4
(d)	6	3
(e)	7	6
(f)	500	400
(g)	0.6×10^3	200
(h)	75	25
(i)	38	19
(j)	52	13

3. A heating element is in two sections, each of $54\,\Omega$ resistance. Calculate the current taken from a 230 V supply when the sections are connected (a) in series, (b) in parallel.

4. Two single-core cables, having resistances of $1.2\,\Omega$ and $0.16\,\Omega$, are connected in parallel and are used to carry a total current of 30 A. Calculate (a) the voltage drop along the cables, (b) the actual current carried by each cable.

5. A cable carries a current of 65 A with a 13 V drop. What must be the resistance of a cable which, when connected in parallel with the first cable, will reduce the voltage drop to 5 V?

6. To vary the speed of a d.c. series motor it is usual to connect a diverter resistor in parallel with the field winding.

   The field of a series motor has a resistance of $0.6\,\Omega$ and the diverter resistor has three steps, of $5\,\Omega$, $4\,\Omega$, and $2\,\Omega$. Assuming that the total current is fixed at 28 A, find out how much current flows through the field winding at each step of the diverter.

7. Resistors of $24\,\Omega$ and $30\,\Omega$ are connected in parallel. What would be the value of a third resistor to reduce the combined resistance to $6\,\Omega$?

8. Two cables having resistances of $0.03\,\Omega$ and $0.04\,\Omega$ between them carry a total current of 70 A. How much does each carry?

9. When two equal resistors are connected in series to a 125 V supply, a current of 5 A flows. Calculate the total current which would flow from the same voltage supply if the resistors were connected in parallel.

10. A current of 50 A is carried by two cables in parallel. One cable has a resistance of $0.15\,\Omega$ and carries 20 A. What is the resistance of the other cable?

11. Three cables, having resistances of $0.018\,\Omega$, $0.024\,\Omega$, and $0.09\,\Omega$ respectively, are connected in parallel to carry a total current of 130 A. Calculate
    (a) the equivalent resistance of the three in parallel,
    (b) the voltage drop along the cables,
    (c) the actual current carried by each cable.

12. Four resistance coils – A, B, C, and D – of values $4\,\Omega$, $5\,\Omega$, $6\,\Omega$, and $7\,\Omega$ respectively, are joined to form a closed circuit in the form of a square. A direct-current supply at 40 V is connected across the ends of coil C. Calculate
    (a) the current flowing in each resistor,
    (b) the total current from the supply,
    (c) the potential difference across each coil,
    (d) the total current from the supply if a further resistance coil R of $8\,\Omega$ is connected in parallel with coil A.

13. Resistors of $3\,\Omega$, $5\,\Omega$, and $8\,\Omega$ are connected in parallel. Their combined resistance is
    (a) $1.6\,\Omega$ (b) $0.658\,\Omega$ (c) $16.0\,\Omega$ (d) $1.52\,\Omega$

14. Two resistors are connected in parallel to give a combined resistance of $3.5\,\Omega$. The value of one resistor is $6\,\Omega$. The value of the other is
    (a) $8.4\,\Omega$ (b) $0.12\,\Omega$ (c) $1.2\,\Omega$ (d) $2.5\,\Omega$

15. The resistance of a cable carrying 43 A is $0.17\,\Omega$. Calculate the resistance of a second cable which, if connected in parallel, would reduce the voltage drop to 5 V.

16. A cable of resistance $1.92\,\Omega$ carries a current of 12.5 A. Find the voltage drop. If a second cable of $2.04\,\Omega$ resistance is connected in parallel, what voltage drop will occur for the same value of load current?

17. Three cables, having resistances $0.0685\,\Omega$, $0.0217\,\Omega$, and $0.1213\,\Omega$, are connected in parallel. Find (a) the resistance of the combination, (b) the total current which could be carried by the cables for a voltage drop of 5.8 V.

18. A load current of 250 A is carried by two cables in parallel. If their resistances are $0.0354\,\Omega$ and $0.046\,\Omega$, how much current flows in each cable?

19. Two cables in parallel between them carry a current of 87.4 A. One of them has a resistance of 0.089 Ω and carries 53 A. What is the resistance of the other?

20. Resistors of 34.7 Ω and 43.9 Ω are connected in parallel. Determine the value of a third resistor which will reduce the combined resistance to 19 Ω.

21. Three pvc-insulated cables are connected in parallel, and their resistances are 0.012 Ω, 0.015 Ω, and 0.008 Ω respectively. With a total current of 500 A flowing on a 240 V supply,

    (a) calculate the current in each cable;
    (b) calculate the combined voltage drop over the three cables in parallel;
    (c) calculate the individual voltage drop over each cable in the paralleled circuit.

22. Tests on a 300 m length of single-core mineral insulated cable produced the following results: conductor resistance 2.4 Ω, insulation resistance 40 MΩ. What will be the anticipated conductor and insulation resistance values of a 120 m length of the cable?

    (a) 16 Ω, 0.96 MΩ   (c) 0.96 Ω, 40 MΩ
    (b) 0.96 Ω, 16 MΩ   (d) 16 Ω, 16 MΩ

23. A 250 m reel of twin mineral insulated cables is to be cut to provide two equal lengths. Before cutting the cable one core is tested and the insulation resistance is found to be 23 MΩ and the conductor resistance found to be 2.9 Ω. What will be the anticipated conductor and insulation resistance values of each of the two lengths?

    (a) 46 Ω, 1.45 MΩ   (c) 0.145 Ω, 11.5 MΩ
    (b) 1.45 Ω, 46 MΩ   (d) 11.5 Ω, 46 MΩ

SERIES AND PARALLEL RESISTOR CIRCUITS

EXAMPLE Resistors of 4 Ω and 5 Ω are connected in parallel and a 6 Ω resistor is connected in series with the group. The combination is wired to a 100 V supply (Fig. 8). Determine

(a) the total resistance,
(b) the current in each resistor.

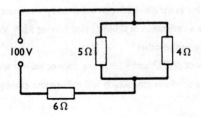

Fig. 8

(a) To find a resistance equivalent to the parallel group,

$$\frac{1}{R} = \frac{1}{R_1} + \frac{1}{R_2}$$

$$= \frac{1}{4} + \frac{1}{5}$$

$$= \frac{5+4}{20}$$

$$= \frac{9}{20}$$

$$\therefore \quad R = \frac{20}{9} = 2.22\,\Omega$$

The circuit may now be redrawn as shown in Fig. 9.

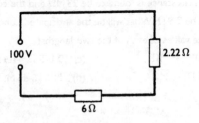

Fig. 9

The total resistance R_t is given by

$$R_t = 2.22 + 6$$

$$= 8.22\,\Omega$$

(b) To find the total current,

$$U = I \times R_t$$

where U is the supply voltage

and R_t is the total resistance

$\therefore$ $100 = I \times 8.22$

$\therefore$ $I = \dfrac{100}{8.22}$

            $= 12.17\,\text{A}$ (the current in the $6\,\Omega$ resistor)

To find the voltage drop in the $6\,\Omega$ resistor,

$$U = I \times R$$

$$= 12.17 \times 6$$

$$= 73.02\,\text{V}$$

$\therefore$ voltage drop across $= 100 - 73.02$
the parallel group

$$= 100 - 73 \quad \text{(ignoring the 0.02)}$$

$$= 27\,\text{V}$$

To find the current through the $4\,\Omega$ and $5\,\Omega$ resistors, use the equation

$$U = I \times R \quad \text{(in this case } U = 27\,\text{V)}$$

When $R = 4$, $27 = I \times 4$

$\therefore$ $I = \dfrac{27}{4}$

                   $= 6.75\,\text{A}$

When $R = 5$, $I = \dfrac{27}{5}$

                   $= 5.4\,\text{A}$

Thus,   current through $4\,\Omega$ resistor $= 6.75\,\text{A}$

        current through $5\,\Omega$ resistor $= 5.4\,\text{A}$

and, by adding as a check,

        current through $6\,\Omega$ resistor $= 12.15\,\text{A}$

INTERNAL RESISTANCE

EXAMPLE A heating element of $2.4\,\Omega$ resistance is connected to a battery of e.m.f. $12\,V$ and internal resistance $0.1\,\Omega$ (Fig. 10). Calculate

(a) the current flowing,
(b) the terminal voltage of the battery on load,
(c) the power dissipated by the heater.

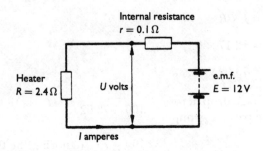

Fig. 10

(a) Using E for e.m.f. and U for terminal voltage, and treating the internal resistance as an additional series resistance,

$$\text{e.m.f.} = \text{total current} \times \text{total resistance}$$

$$E = I \times (R + r) \quad \text{(note the use of brackets)}$$

$$\therefore \quad 12 = I \times (2.4 + 0.1)$$

$$= I \times 2.5$$

$$\therefore \quad I = \frac{12}{2.5}$$

$$= 4.8\,A$$

(b) The terminal voltage is the e.m.f. minus the voltage drop across the internal resistance:

$$\text{terminal voltage } U = E - Ir$$

$$= 12 - (4.8 \times 0.1)$$

$$= 12 - 0.48$$

$$= 11.52 \quad \text{or} \quad 11.5\,V$$

(c) The power dissipated in the heater is

$$P = U \times I$$
$$= 11.5 \times 4.8$$
$$= 55\,\text{W}$$

EXERCISE 4

1. For the circuit of Fig. 11, find
   (a) the resistance of the parallel group,
   (b) the total resistance,
   (c) the current in each resistor.

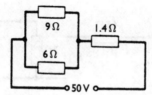

Fig. 11

2. For the circuit of Fig. 12, find
   (a) the total resistance,
   (b) the supply voltage.

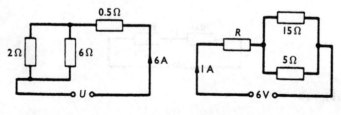

Fig. 12

Fig. 13

3. Find the value of the resistor R in the circuit of Fig. 13.
4. Calculate the value of the resistor r in the circuit of Fig. 14.

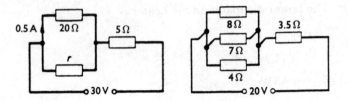

Fig. 14

Fig. 15

5. For the circuit of Fig. 15, find
   (a) the total resistance,
   (b) the total current.
6. Determine the voltage drop across the 4.5 Ω resistor in Fig. 16.

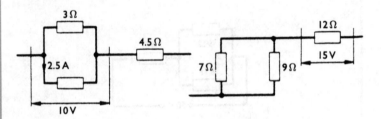

Fig. 16

Fig. 17

7. Calculate the current in each resistor in Fig. 17.
8. Determine the value of a resistor which when connected in parallel with
   the 70 Ω resistor will cause a total current of 2.4 A to flow in the circuit
   of Fig. 18.

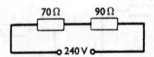

Fig. 18

9. Two contactor coils of resistance 350 Ω and 420 Ω respectively are
   connected in parallel. A ballast resistor of 500 Ω is connected in series
   with the pair. Supply is taken from a 220 V d.c. supply. Calculate the
   current in each coil and the power wasted in the ballast resistor.

10. Two 110 V lamps are connected in parallel. Their ratings are 150 W and 200 W. Determine the value of a resistor which when wired in series with the lamps will enable them to operate from the 230 V mains.

11. A shunt motor has two field coils connected in parallel, each having a resistance of 235 Ω. A regulating resistor is wired in series with the coils to a 200 V supply. Calculate the value of this resistor when the current through each coil is 0.7 A.

12. In a certain installation the following items of equipment are operating at the same time: (i) a 3 kW immersion heater, (ii) two 100 W lamps, (iii) one 2 kW radiator. All these are rated at 240 V.

    The nominal supply voltage is stated to be 230 V but it is found that the actual voltage at the origin of the installation is 5 V less than this. Calculate

    (a) the total current,

    (b) the resistance of the supply cables,

    (c) the actual power absorbed by the immersion heater.

13. The overhead cable supplying an outbuilding from the 230 V mains supply has a resistance of 0.9 Ω. A 2 kW radiator and a 1500 W kettle, both rated at 230 V, are in use at the time. Determine the voltage at the terminals of this apparatus. What would be the voltage if a 750 W, 240 V water heater were also switched on?

14. Two resistors in parallel, A of 20 Ω and B of unknown value, are connected in series with a third resistor C of 12 Ω. The supply to the circuit is direct current.

    If the potential difference across the ends of C is 180 V and the power in the complete circuit is 3600 W, calculate

    (a) the value of resistor B,

    (b) the current in each resistor,

    (c) the circuit voltage.

15. State Ohm's law in your own words, and express it in symbols. A d.c. supply at 240 V is applied to a circuit comprising two resistors A and B in parallel, of 5 Ω and 7.5 Ω respectively, in series with a third resistor C of 30 Ω.

    Calculate the value of a fourth resistor D to be connected in parallel with C so that the total power in the circuit shall be 7.2 kW.

16. Three resistors, of value 1.5 Ω, 4 Ω, and 12 Ω respectively, are connected in parallel. A fourth resistor, of 6 Ω, is connected in series

with the parallel group. A d.c. supply of 140 V is applied to the circuit.

   (a) Calculate the current taken from the supply.

   (b) Find the value of a further resistor to be connected in parallel with the 6 Ω resistor so that the potential difference across it shall be 84 V.

   (c) What current will now flow in the circuit?

17. An electric bell takes a current of 0.3 A from a battery whose e.m.f. is 3 V and internal resistance 0.12 Ω. Calculate the terminal voltage of the battery when the bell is ringing.

18. Determine the voltage at the terminals of a battery of three cells in series, each cell having an e.m.f. of 1.5 V and internal resistance 0.11 Ω, when it supplies a current of 0.75 A.

19. A car battery consists of six cells connected in series. Each cell has an e.m.f. of 2 V and internal resistance of 0.008 Ω. Calculate the terminal voltage of the battery when a current of 105 A flows.

20. A battery has an open-circuit voltage of 6 V. Determine its internal resistance if a load current of 54 A reduces its terminal voltage to 4.35 V.

21. Resistors of 5 Ω and 7 Ω are connected in parallel to the terminals of a battery of e.m.f. 6 V and internal resistance of 0.3 Ω. Calculate

   (a) the current in each resistor,

   (b) the terminal voltage of the battery,

   (c) the power wasted in internal resistance.

22. A battery is connected to two resistors, of 20 Ω and 30 Ω, which are wired in parallel. The battery consists of three cells in series, each cell having an e.m.f. of 1.5 V and internal resistance 0.12 Ω. Calculate

   (a) the terminal voltage of the battery,

   (b) the power in each resistor.

23. A battery of 50 cells is installed for a temporary lighting supply. The e.m.f. of each cell is 2 V and the internal resistance is 0.0082 Ω. Determine the terminal voltage of the battery when it supplies 25 lamps each rated at 150 W, 110 V.

24. The installation in a country house is supplied from batteries. The batteries have an open-circuit voltage of 110 V and an internal resistance of 0.045 Ω. The main cables from the batteries to the house have a resistance of 0.024 Ω. At a certain instant the load consists of two 2 kW

radiators, three 100 W lamps, and four 60 W lamps. All this equipment is rated at 110 V. Calculate the voltage at the apparatus terminals.

25. An installation is supplied from a battery through two cables in parallel. One cable has a resistance of $0.34\,\Omega$; the other has a resistance of $0.17\,\Omega$. The battery has an internal resistance of $0.052\,\Omega$ and its open-circuit voltage is 120 V. Determine the terminal voltage of the battery and the power wasted in each cable when a current of 60 A is flowing.

26. A 12 V battery needs charging and the only supply available is one of 24 V. The battery has six cells, each of e.m.f. 1.8 V and internal resistance $0.009\,\Omega$. Determine the value of a series resistor which will limit the current to 5 A.

27. A circuit consists of a $7.2\,\Omega$ resistor in parallel with one of unknown value. This combination is connected in series with a $4.5\,\Omega$ resistor to a supply of direct current. The current flowing is 2.2 A and the total power taken by the circuit is 35 W. Calculate

(a) the value of the unknown resistor,

(b) the supply voltage,

(c) the value of a resistor which if connected in parallel with the $4.5\,\Omega$ resistor will cause a current of 4 A to flow.

(Assume that the source of supply has negligible internal resistance.)

28. The combined resistance of the circuit in Fig. 19 is

(a) $0.333\,\Omega$ (b) $12.5\,\Omega$ (c) $30.0\,\Omega$ (d) $7.7\,\Omega$

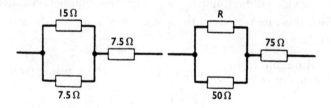

Fig. 19   **Fig. 20**

29. The combined resistance of the circuit in Fig. 20 is $91.7\,\Omega$. The value of resistor R is

(a) $33.3\,\Omega$ (b) $250\,\Omega$ (c) $0.04\,\Omega$ (d) $25\,\Omega$

30. The current flowing in the $0.4\,\Omega$ resistor in Fig. 21 is

(a) 8.57 A   (b) 11.43 A   (c) 0.24 A   (d) 0.73 A

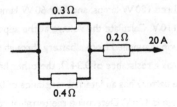

Fig. 21

$$\text{Fusing factor} = \frac{\text{minimum current required to blow the fuse}}{\text{rated current of the fuse}}$$

EXAMPLE If a semi-enclosed fuse (BS 3036) requires about 1.8 times its nominal current (I_n) to cause it to blow (fusing factor), calculate

(a) the current required to blow a 15 A fuse,

(b) the value of fault impedance of the live conductors above which this fuse would fail to blow at normal mains voltage of 230 V.

(a) Minimum fusing current (I_2) $=$ nominal current (I_n) $\times$ fusing factor

$$= 15 \times 1.8$$

$$= 27\,\text{A}$$

(b) Figure 22 illustrates the path taken by the fault current (I_f) during phase-to-neutral-conductor fault conditions.
The maximum impedance of the fault path to allow the fuse to operate may be found by Ohm's law as follows:

$$U_0 = I_2 \times Z_f$$

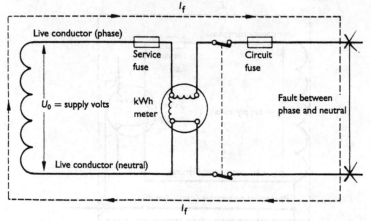

Fig. 22

where U_0 = voltage at the source (supply transformer)

I_2 = operating current of the device (fuse)

and Z_f = impedance of the fault path

$\therefore$ $230 = 27 \times Z_f$

$\therefore$ $\dfrac{230}{27} = Z_f$

$\therefore$ $Z_f = 8.5\,\Omega$

If the impedance of the fault path is greater than this, insufficient current will flow and the fuse will fail to blow.

EARTH LEAKAGE CURRENT PROTECTION

Figure 23 illustrates the path which is taken by the earth leakage current during a fault between a phase conductor and an exposed conductive part at the origin of an installation. This current could cause shock risk between the metalwork of the premises and the general mass of earth if the protective device (the service fuse) does not operate quickly and disconnect the installation from the supply.

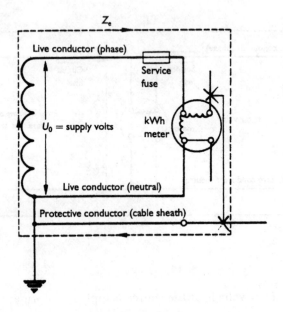

Fig. 23

The earth leakage current may be determined by applying the modified Ohm's law formula:

$$I_f = \frac{U_0}{Z_e}$$

where I_f = earth leakage current

  U_0 = voltage at source (phase to earthed neutral)

and Z_e = impedance of external earth leakage loop

EXAMPLE The 100 A H.B.C. service fuse (BS 1361) at the origin of an installation has a fusing factor of 1.4, the voltage at the supply transformer (U_0) is 230 V, and the tested value of Z_e at the origin of the installation is $0.36\,\Omega$.

(a) How much current will be required to blow the fuse?

(b) What earth leakage current will flow if the phase conductor comes into direct contact with the protective conductor (the service-cable sheath) at the origin of the installation?

(c) Will the fuse operate under the conditions of (b)?

(a) Minimum fusing current $(I_2) = 100 \times 1.4$

$$= 140\,\text{A}$$

(b) Earth leakage current $(I_f) = \dfrac{U_0}{Z_e}$

$$= \dfrac{230}{0.36}$$

$$= 638.9\,\text{A}$$

(c) The fuse will therefore operate. (The speed of operation may be estimated by reference to the characteristic graphs contained in Appendix 3 of British Standard 7671.)

From Fig. 3.1, Appendix 3, the disconnection time is approximately 3.8 s.

ELECTRIC SHOCK PROTECTION

In the previous example, the earth leakage current was over twice that required to operate (blow) the fuse. However, the consumer's circuits increase the earth-loop impedance – i.e. the phase conductor will have resistance (R_1) as will the protective conductor (R_2) – and thus the total earth-loop impedance $(Z_S = Z_e + R_1 + R_2)$ may cause the earth leakage current to be reduced to a value which will not operate the fuse. If the fuse does not operate rapidly, a shock voltage may exist between the *exposed conductive parts* of the installation and *extraneous conductive parts* of the premises (the metal conduit and metal water or gas pipes) during the time it takes the fuse to operate.

EXAMPLE Immediately following the service fuse and meter in the installation described in the previous example there is a switch fuse containing a 30 A BS 3036 semi-enclosed fuse to protect a radial circuit feeding stationary equipment. The fuse has a fusing factor of 1.8. The resistance of the phase conductor to the fixed equipment (R_1) is $0.05\,\Omega$ and, due to a damaged conduit joint, the resistance of the protective conductor (R_2) is $6\,\Omega$.

(a) Calculate

(i) the value of the total earth-loop impedance (Z_S),

(ii) the prospective earth-fault current (I_f),

(iii) the prospective shock voltage between the metal case of the fixed equipment and the general mass of earth.

(b) Will the fuse operate and render the circuit safe?

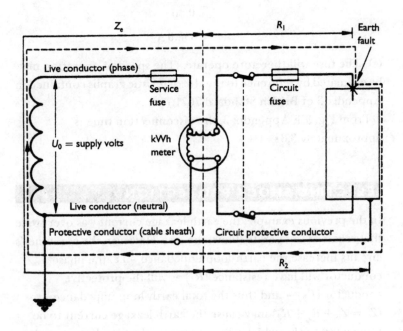

Fig. 24

(a) Figure 24 illustrates the path of the earth leakage current during the fault.

(i) From the previous example, the value of Z_e is $0.36\,\Omega$.

Now $Z_S = Z_e + R_1 + R_2$

$\therefore \qquad Z_S = 0.36 + 0.05 + 6$

$\qquad\qquad = 6.41\,\Omega$ (see note on p. 33).

(ii) $\quad I_f = \dfrac{230}{6.41}$

$\qquad\qquad = 35.88\,\text{A}$

(iii) Prospective shock voltage

$$U_f = 35.88 \times 6$$

$$= 215.28\,\text{V}$$

(b) As the fusing factor is 1.8, it will require $30 \times 1.8 = 54\,\text{A}$ to operate the fuse, so in this case the fuse will not operate and the shock danger will remain for an indefinite time.

DISCONNECTION TIMES

BS 7671 Requirements Part 4, Chapters 41 and 47 lay down maximum disconnection times for circuits under earth fault conditions.

Maximum disconnection time for 230 V circuits feeding only stationary equipment is 5 seconds (Regulation 413–02–13). Table 41D gives guidance to the maximum total earth fault loop impedance values (Z_S) to achieve this disconnection time.

Note The circuit in the above example is feeding stationary equipment, so if table 41D is consulted this states that the maximum value of Z_S for a 30 A BS 3036 fuse should not exceed 2.76 Ω.

Maximum disconnection times for 230 V socket outlet circuits and other 230 V final circuits which supply portable equipment should not exceed 0.4 s (Regulation 413–02–09 and table 41A).

Tables 41B1 and 41B2 may be used to establish the maximum earth fault loop impedance Z_S to achieve a 0.4 s disconnection time.

EXERCISE 5

1. Complete the following table:

U (volts)	10	20		40	
I (amperes)	1		3	4	5
R (ohms)		10	10		10

2. Using the values from question 1, correctly completed, plot a graph of current against voltage. Take 1 cm ≡ 1 A vertically and 1 cm ≡ 10 V horizontally.

Read from the graph the voltage required to produce a current of 3.6 A.

3. Complete the following table:

U (volts)		240			240	
I (amperes)	12	6	4	3	2.4
R (ohms)	20		60		100

4. When the table of question 3 has been correctly completed, plot a graph showing the relationship between current and resistance. Use the graph to find the value of the current when the resistance is 78 Ω. State also the value of resistance required to give a current of 9.5 A.

5. Complete the following table:

U (volts)	100	100		56	96	132	84	144		
I (amperes)	10		12	7	8	12		12	11	9
R (ohms)		10	8				12		11	7

6. Complete the following table:

I (amperes)	100		10		0.1	0.1		0.001	0.1	
R (ohms)	0.1	1000	0.1	1000	0.1		0.1			0.01
U (volts)		100		10		100	10	20	200	2

7. Complete the following table;

R (ohms)		14		16		0.07	12			15
I (amperes)	0.5	15	0.05		1.2		0.9		0.2	
U (volts)	240		25	96	132	8.4		100	6	120

8. A cable of resistance 0.029 Ω carries a current of 83 A. What will be the voltage drop?

9. To comply with a BS 7671 regulation, the maximum value of voltage drop which can be tolerated in a circuit supplied from the 230 V mains is 9.2 V. Calculate the maximum resistance which can be allowed for circuits carrying the following currents:

(a) 28 A   (b) 53 A   (c) 77 A   (d) 13 A   (e) 203 A

10. The cable in a circuit has a resistance of 0.528 Ω. What is the maximum current it can carry if the voltage drop is not to exceed 5.8 V?

11. A 50 V a.c. system supplies the following loads by means of a radial circuit:

load A: 15 A at a distance of 18 m from the supply point S,

load B: 25 A at a distance of 35 m from A,

load C: 20 A at a distance of 43 m from B.

The type of cable used produces a voltage drop of 2.7 mV per ampere per metre.

Calculate the voltage drop in each section of the circuit and the voltage at each load point.

12. Assuming a fusing factor of 1.4, complete the following table, which refers to various sizes of fuse:

Nominal current (A)	5	15	30	60	100
Minimum fusing current (A)					

13. Repeat exercise 12 using a fusing factor of 1.2.

14. Refer to Fig. 24 on page 32 to calculate the voltage between the metal parts and earth under the following fault conditions. The supply voltage in each case is 230 V.

	Rating of circuit fuse (I_n) (A)	Fusing factor	Resistance of CPC (R_2) etc. (Ω)	Resistance of remainder of loop $(Z_e + R_1)$ (Ω)
(a)	30	1.3	5	0.5
(b)	30	1.2	8	0.5
(c)	30	1.2	0.5	7
(d)	60	1.5	4	1
(e)	100	1.2	4	1
(f)	100	1.2	2.5	0.5

15. A current of 1.5 A flows in a 25 Ω resistor. The voltage drop is
    (a) 0.06 V   (b) 37.5 V   (c) 16.67 V   (d) 3.75 V

16. If a cable must carry a current of 19.5 A with a voltage drop of not more than 6 V, its resistance must not exceed
    (a) 32.5 Ω (b) 117 Ω (c) 0.308 Ω (d) 3.25 Ω

17. A fuse rated at 30 A has a fusing factor of 1.4. The current required to blow the fuse is
    (a) 31.4 A   (b) 21.4 A   (c) 42 A   (d) 30 A

18. A faulty earthing conductor has a resistance of 12.5 Ω, and the resistance of the remainder of the fault path is 1.5 Ω. The supply voltage is 230 V. The voltage appearing between metal parts and earth is
    (a) 205.4 V   (b) 238.5 V   (c) 24.6 V   (d) 217.7 V

DIRECT AND INDIRECT PROPORTION

EXAMPLE 1 If a heater takes 12 A from a 230 V supply, what current will it take if the voltage falls to 220 V?

The *lower* voltage will give *less* current,

$$\therefore \quad \text{current at } 220\,\text{V} = 12 \times \frac{220\ (smaller)}{230\ (larger)}$$

$$= 11.4\,\text{A}$$

EXAMPLE 2 An electric sign takes 25 A from a 230 V supply. What current will it take if the voltage is raised to 240 V?

The *higher* voltage will cause *more* current to flow,

$$\therefore \quad \text{current at } 240\,\text{V} = 25 \times \frac{240(larger)}{230(smaller)}$$

$$= 26.1\,\text{A} \quad \text{(to three significant figures)}$$

EXAMPLE 3 The current through a heating element is 5 A when the voltage is 100 V. What voltage must be applied to obtain a current of 4.5 A?

The *smaller* current will require a *lower* voltage,

$$\therefore \quad \text{voltage to produce } 4.5\,\text{A} = 100 \times \frac{4.5}{5}$$

$$= 90\,\text{V}$$

EXAMPLE 4 The current through the field coils of a motor is 2 A when the resistance is 250 Ω. Due to rise in temperature, the resistance increases to 275 Ω. If the voltage remains the same, what will be the value of the current?

Greater resistance gives *less* current,

$$\therefore \quad \text{new current} = 2 \times \frac{250}{275}$$

$$= 1.82\,\text{A}$$

EXAMPLE 5 A heating element having a resistance of 54 Ω takes a current of 4.5 A. What must its resistance be for it to take 5 A at the same voltage?

A *larger* current requires *less* resistance,

$$\therefore \quad \text{new resistance} = 54 \times \frac{4.5}{5}$$

$$= 48.6\,\Omega$$

EXERCISE 6

1. A contactor coil takes a current of 0.6 A when its impedance is 400 Ω. Find by the method of proportion what current it will take when its impedance is
   (a) 450 Ω (b) 380 Ω (c) 426 Ω (d) 392 Ω (e) 405 Ω.

2. A resistor is variable in steps of 15 Ω from 95 Ω down to 20 Ω. If it takes 75 mA at the 80 Ω step, find by proportion what current it will take at each step from the same voltage supply.

3. A heating element arranged for the three-heat control takes 4.8 A when its resistance is 52 Ω. Determine the current which flows when the connections are such that its resistance is
   (a) 13 Ω   (b) 26 Ω.

4. The resistance of the shunt field circuit of a motor is 360 Ω and the current through it is 1.2 A. Determine the value of an additional series resistor which will reduce the current to 1.05 A.

5. The coil of a relay takes 0.13 A from a 24 V supply. Calculate by proportion the voltage necessary to produce a current of
   (a) 0.15 A (b) 93 mA (c) 0.095 A (d) 80 mA (e) 0.125 A.

6. Complete the following table:

U (volts)		40	101	80	120	60
I (amperes)	0.2		0.2		0.195	
Z (ohms)	100	200		400		300

   Plot a graph showing the relationship between voltage and impedance. From the graph, state the voltage when the impedance is
   (a) 150 Ω   (b) 320 Ω.

CURRENT RATING OF CABLES AND VOLTAGE DROP

British Standard 7671:1992 and its supporting documents are intended for use by all those persons employed in designing and actually carrying out the practical process of electrical installation. Obviously some of these persons, e.g. designers of large factory or commercial installations, will require a deeper and wider knowledge of BS 7671 than those simply carrying out small domestic-type installations.

It is appropriate therefore for this book to concentrate upon the basic methods of calculating and selecting cables for the smaller installation and leaving more complex cable calculations to a later book. For the purposes of this book the full BS 7671 document is not necessary and 'site'-type data books such as the IEE On-Site Guide or similar Electricians' guides will satisfy the basic calculations.

Application of correction factors

The first step in calculating the minimum current-carrying capacity or current rating of a circuit cable in accordance with BS 7671 is to establish the design current of the circuit (I_b), then select an overcurrent protective device in accordance with BS 7671 regulation 433–02.

The minimum cable current rating (I_t) is then found by the use of the formula

$$I_t = \frac{I_n}{C_a C_i C_g C_r}$$

where I_n is the current rating of the fuse or the setting of the circuit breaker protecting the circuit, and:

C_a is a correction factor for ambient temperature (table 6A1 and 6A2 in the IEE On-Site Guide)

C_i is a correction factor for thermal insulation (table 6B in the IEE On-Site Guide)

C_g is a correction factor for cable grouping (table 6C in the IEE On-Site Guide)

C_r is a correction factor used when semi-enclosed fuses to BS 3036 (re-wireable fuses) are used.

Not all factors may be necessary, but care must be taken to identify the actual conditions which will exist in each of the areas the cable passes through. It is often possible to determine a cable rating based simply upon the most hostile environmental area and accept it as the minimum cable to be used throughout the run.

The correction factor C_r (0.725) however must always be used when semi-enclosed fuses protect a circuit wired in pvc-insulated cables; where mineral-insulated cables are used then C_r may be disregarded.

Voltage drop calculation

The voltage drop in a cable is directly proportional to the circuit current and the length of cable run:

$$\text{Voltage drop} = \frac{\text{current (A)} \times \text{length of run (m)} \times \text{millivolt drop per A/m}}{1000}$$

(Note: the division by 1000 to convert millivolts to volts.)

Note BS 7671 Regulation 525-01-02 limits the voltage drop permitted between the origin of the installation and the terminals of a load to 4% of the nominal supply voltage; for a 230 V supply this equates to 9.2 V.

EXAMPLE I   A domestic immersion heater drawing 13 A is to be wired using pvc-insulated twin and earth cable. The length of run is 15 m, of which 6 m is within the empty roof space of the building. Protection is by a BS 3036 fuse within a consumer unit. Assume that the voltage drop in the circuit is limited to 8.5 V.

(a) Calculate

(i)   the minimum current carrying capacity of the cable,

(ii)   the minimum cross-sectional area of the cable, and

(iii) the voltage drop in the cable.

(b) What consideration should be given to the cable run in the roof space?

(a) As there are at present no obvious adverse environmental conditions, it being a domestic premises, we need not at this point apply correction factors C_a, C_i or C_g but we must apply the factor C_r of 0.725 because BS 3036 fuses are used.

Select a 15 A BS 3036 fuse as I_n.

(i) So $I_t = \dfrac{15}{0.725}$

$\qquad = 20.7\,\text{A}$

Using the IEE On-Site Guide:

(ii) Select from table 6E1, column 6, 2.5 mm^2 cable (27 A).

(iii) From table 6E2 the mV/A/m value for 2.5 mm^2 two-core cable is 18 mV/A/m.

So voltage drop in 15 m $= \dfrac{13 \times 15 \times 18}{1000}$

$\qquad\qquad\qquad\qquad = 3.51\,\text{V}$

This being less than 8.5 V the cable is satisfactory.

(b) The electrician installing the cable should make every effort to assess whether the cable will be likely to be covered or surrounded by thermal insulation. Cables should wherever practicable be fixed in a position not likely to be so affected. Cables subjected to enclosure in thermal insulation may need to be de-rated to half their normal current-carrying capacity.

EXAMPLE 2 An electric heater has a design current of 40 A and is sited 25 m from its BS 3036 switch-fuse. The circuit is to be wired in pvc-insulated single-core cables enclosed in their own steel conduit. Calculate (a) the minimum current-carrying capacity of the cables, (b) the minimum cross-sectional area of the cables, (c) the voltage drop in the circuit and (d) the effect on (a), (b) and (c) of replacing the BS 3036 switch-fuse by a BS 1361 switch-fuse.

(a) Minimum cable carrying capacity

$$I_t = \frac{I_n}{C_a C_i C_g C_r}$$

In this case only C_r applies.

Assume a 45 A fuse in the switch-fuse

$$\therefore \quad I_t = \frac{45}{0.725}$$

$$= 62\,A$$

Using the IEE On-Site Guide:
(b) From table 6D1 column 4, select 16 mm^2 cables (76 A).
(c) From table 6D2 column 3, the mV/A/m value for 16 mm^2 cable is 2.8.

$$\text{So voltage drop in 25 m} = \frac{40 \times 25 \times 2.8}{1000}$$

$$= 2.8\,V$$

(d) If the BS 3036 switch-fuse was replaced by a BS 1361 switch-fuse then no correction factors are necessary.

$$\therefore \quad I_t = 45\,A$$

Using the IEE On-Site Guide:
From table 6D1, column 4, select 10 mm^2 cables (67 A).
From table 6D2, column 3, the mV/A/m value for 10 mm^2 cable is 4.4.

$$\text{So voltage drop in 25 m} = \frac{40 \times 25 \times 4.4}{1000}$$

$$= 4.4\,V$$

EXAMPLE 3 A single-phase circuit wired in pvc-insulated single-core cables has a design current of 26 A and is protected by a BS 3036 fuse in a distribution board; the circuit cables share a common conduit with three other single-phase circuits. The ambient temperature may be taken as 35 °C. The length of circuit cables is 30 m.

With the aid of tables based on BS 7671 establish the
(a) minimum current carrying capacity of the cables,
(b) minimum cable cross-sectional area (c.s.a.),
(c) voltage drop in the circuit cables.

The design current is 26 A, thus choose a BS 3036 30 A fuse.

Using the IEE On-Site Guide:

Table 6C correction factor (C_g) for four circuits is 0.65.

Table 6A2 correction factor (C_t) for 35 °C is 0.97.

Correction factor (C_r) for BS 3036 fuse is 0.725.

(a) $\quad$ Minimum current carrying $\quad = \dfrac{I_n}{C_g C_t C_r}$
capacity of cables I_t

$$\therefore \quad I_t = \frac{30}{0.65 \times 0.97 \times 0.725}$$

$$= 65.63 \,\text{A}$$

(b) From table 6D1, columns 1 and 4, select $16\,\text{mm}^2$ (76 A).

(c) From table 6D2, column 3, $16\,\text{mm}^2$ cable has a voltage drop (per ampere per metre) of 2.8 mV.

$$\therefore \quad \text{Voltage drop in } 30\,\text{m} = \frac{26 \times 30 \times 2.8}{1000}$$

$$= 2.18 \,\text{V}$$

EXERCISE 7

1. The voltage drop figure for a certain cable is 44 mV/A/m. Calculate the voltage drop in a 15 m run of this cable when carrying a load of 6 A.

2. The design current of a circuit protected by a BS 1361 fuse is 28 A, the grouping correction factor is 0.8, and the ambient temperature correction factor is 1.04. Calculate the minimum current-carrying capacity of the cable.

3. A circuit is protected by a BS 3871 circuit breaker rated at 30 A. The grouping correction factor is 0.54 and the ambient temperature correction factor is 0.94. Calculate the minimum current capacity of the cable.

4. Calculate the effect on the minimum cable current rating required in question 3 if the circuit breaker is replaced by a BS 3036 semi-enclosed fuse.

5. A cable with a voltage drop figure of 6.4 mV/A/m supplies a current of 24 A to a point 18 m away from a 230 V supply source. Determine (a) the voltage drop in the cable and (b) the actual voltage at the load point.

6. There is a voltage drop limitation of 5 V for a circuit wired in pvc-insulated twin and earth cable (clipped direct), having a length of run of 35 m. The current demand is assessed as 36 A. Protection is by a BS 1361 fuse.

Establish the:

(a) fuse rating,

(b) maximum mV/A/m value,

(c) minimum cable cross-sectional area,

(d) actual voltage drop in the chosen cable.

7. A 230 V heater has a total current demand of 33 A and is to be wired using pvc-insulated single-core cables in pvc conduit.

Assume a temperature correction factor of 0.97 is applicable, and the circuit length is 27 m. Protection is by a BS 3871 circuit breaker. The circuit voltage drop is not to exceed 4 V.

Determine the:

(a) fuse rating,

(b) minimum cable current rating,

(c) minimum cable cross-sectional area,

(d) maximum mV/A/m value,

(e) actual voltage drop.

Power in a d.c. circuit

METHOD 1

Power (watts) = voltage (volts) × current (amperes)

$$P = U \times I$$

EXAMPLE 1 The current in a circuit is 4.8 A when the voltage is 240 V. Calculate the power.

$P = U \times I$

$= 240 \times 4.8$

$= 1152\,\text{W}$

EXAMPLE 2 Calculate the current flowing when a 2 kW heater is connected to a 230 V supply.

$$P = U \times I$$
$$2000 = 230 \times I$$
$$\therefore \quad I = \frac{2000}{230}$$
$$= 8.7\,\text{A}$$

EXAMPLE 3 The current in a certain resistor is 15 A and the power absorbed is 200 W. Find the voltage drop across the resistor.

$$P = U \times I$$
$$200 = U \times 15$$
$$\therefore \quad U = \frac{200}{15}$$
$$= 13.3\,\text{V}$$

EXERCISE 8

1. Complete the following table:

P (watts)		3000	1600	1000		1000	2350	
I (amperes)	6			150	0.2			4.5
U (volts)	240	250	240		100	220	460	240

2. The voltage drop in a cable carrying 12.5 A is 2.4 V. Calculate the power wasted.

3. A d.c. motor takes 9.5 A from a 460 V supply. Calculate the power input to the motor.

4. Calculate the current that flows when each of following pieces of equipment is connected to the 230 V mains:

    (a) 3 kW immersion heater   (f) 60 W lamp
    (b) 1500 W kettle   (g) 100 W lamp
    (c) 450 W electric iron   (h) 2 kW radiator
    (d) 3.5 kW washing machine   (i) 750 W water heater
    (e) 7 kW cooker   (j) 15 W lamp

5. Calculate the voltage drop in a resistor passing a current of 93 A and absorbing 10 kW.

6. A cable carries a current of 35 A with a 5.8 V drop. Calculate the power wasted in the cable.

7. A heater is rated at 4.5 kW, 240 V. Calculate the current it takes from

   (a) a 240 V supply   (b) a 220 V supply.

8. A motor-starting resistor passes a current of 6.5 A and causes a voltage drop of 115 V. Determine the power wasted in the resistor.

9. Determine the current rating of the resistance wire which would be suitable for winding the element of a 1.5 kW, 250 V heater.

10. Calculate the current taken by four 750 W lamps connected in parallel to a 230 V main.

11. A faulty cable joint causes an 11.5 V drop when a current of 55 A is flowing. Calculate the power wasted at the joint.

12. Two lamps, each with a rating of 100 W at 240 V, are connected in series to a 230 V supply. Calculate the current taken and the power absorbed by each lamp.

13. Determine the current rating of the cable required to supply a 4 kW immersion heater from a 230 V mains.

14. A generator delivers a current of 28.5 A through cables having a total resistance of 0.103 Ω. The voltage at the generator terminals is 225 V. Calculate

    (a) the power generated,

    (b) the power wasted in the cables,

    (c) the voltage at the load.

15. Calculate the value of resistance which when connected in series with a 0.3 W, 2.5 V lamp will enable it to work from a 6 V supply.

16. A motor takes a current of 15.5 A at a terminal voltage of 455 V. It is supplied through cables of total resistance 0.32 Ω. Calculate

    (a) the voltage at the supply end,

    (b) the power input to the motor,

    (c) the power wasted in the cables.

17. Two coils, having resistances of 35 Ω and 40 Ω, are connected to a 100 V d.c. supply (a) in series, (b) in parallel. For each case, calculate the power dissipated in each coil.

18. Two cables, having resistances of 0.036 Ω and 0.052 Ω, are connected in parallel to carry a total current of 190 A. Determine the power loss in each cable.

45

19. If the power loss in a resistor is 750 W and the current flowing is 18.5 A, calculate the voltage drop across the resistor. Determine also the value of an additional series resistor which will increase the voltage drop to 55 V when the same value of current is flowing. How much power will now be wasted in the original resistor?

20. A d.c motor takes a current of 36 A from the mains some distance away. The voltage at the supply point is 440 V and the cables have a total resistance of 0.167 Ω. Calculate

    (a) the voltage at the motor terminals,

    (b) the power taken by the motor,

    (c) the power wasted in the cables,

    (d) the voltage at the motor terminals if the current increases to 42 A.

21. The voltage applied to a circuit is 240 V, and the current is 3.8 A. The power is

    (a) 632 W   **(b)** 63.2 W   **(c)** 912 W   **(d)** 0.016 W

22. The power absorbed by a heating element is 590 W at a p.d. of 235 V. The current is

    (a) 13 865 A   **(b)** 2.51 A   **(c)** 0.34 A   **(d)** 25.1 A

23. A faulty cable joint carries a current of 12.5 A, and a voltage drop of 7.5 V appears across the joint. The power wasted at the joint is

    (a) 1.67 W   **(b)** 0.6 W   **(c)** 93.8 W   **(d)** 60 W

24. A heating element absorbs 2.5 kW of power and the current is 10.5 A. The applied voltage is

    (a) 238 V   **(b)** 26.3 V   **(c)** 2.38 V   **(d)** 4.2 V

METHOD 2

$$\text{Power} = \text{current}^2 \times \text{resistance}$$
$$P = I^2 R$$

EXAMPLE I Calculate the power absorbed in a resistor of 8 Ω when a current of 6 A flows.

$$P = I^2 R$$
$$= 6^2 \times 8$$
$$= 36 \times 8$$
$$= 288 \text{ W}$$

EXAMPLE 2 A current of 12 A passes through a resistor of such value that the power absorbed is 50 W. What is the value of this resistor?

$$P = I^2 R$$
$$50 = 12^2 \times R$$
$$\therefore \quad R = \frac{50}{12 \times 12}$$
$$= 0.347 \, \Omega$$

EXAMPLE 3 Determine the value of current which when flowing in a resistor of 400 Ω causes a power loss of 1600 W.

$$P = I^2 R$$
$$\therefore \quad 1600 = I^2 \times 400$$
$$\therefore \quad I^2 = \frac{1600}{400} = 4$$
$$\therefore \quad I = \sqrt{4} = 2 \, A$$

EXERCISE 9

1. Complete the following table:

Power (W)		200		1440	1000	2640	100	
Current (A)	10	5	15			4.2		0.42	1.3
Resistance (Ω)	15		8	10			20		25

2. A current of 20 A flows in cable of resistance 0.325 Ω. Calculate the power loss.

3. Determine the power loss in a cable having a resistance of 0.14 Ω when passing a current of 14.5 A.

4. Determine the value of current which, when flowing in a 40 Ω resistor, dissipates 1000 W.

5. An earth fault current of 38 A passes through a conduit joint which has a resistance of 1.2 Ω. Calculate the power dissipated at the joint.

6. A 100 W lamp passes a current of 0.42 A. Calculate its resistance.

7. In a certain installation the *total* length of cable is 90 m and the resistance of this type of cable is 0.6 Ω per 100 m. Determine (a) the voltage drop, (b) the power loss, when a current of 11.5 A flows.

8. A resistor used for starting a d.c. motor has a value of 7.5 Ω. Calculate the power wasted in this resistor when a starting current of 8.4 A flows.

9. Determine the current rating of resistance wire which would be suitable for a 1000 W heater element of resistance 2.5 Ω.

10. An ammeter shunt carries a current of 250 A and its resistance is 0.000 95 Ω. Calculate the power absorbed by the shunt.

11. What is the resistance of an electric-iron element of 450 W rating and which takes a current of 1.9 A?

12. A joint in a cable has a resistance of 0.045 Ω. Calculate the power wasted at this joint when a current of 37.5 A flows.

13. The resistance measured between the brushes of a d.c. motor is 2.3 Ω. Calculate the power loss in the armature when the current is 13.5 A.

14. Determine the rating in watts of a 1100 Ω resistor which will carry 15 mA.

15. Calculate the maximum current which a 250 Ω resistor rated at 160 W will carry.

METHOD 3

$$\text{Power} = \frac{\text{voltage}^2}{\text{resistance}}$$
$$P = \frac{U^2}{R}$$

EXAMPLE 1 Calculate the power absorbed by a 40 Ω resistor when connected to a 240 V d.c. supply

$$\text{Power absorbed } P = \frac{U^2}{R}$$
$$= \frac{240 \times 240}{40}$$
$$= 1440 \text{ W}$$

EXAMPLE 2 Determine the resistance of a heater which absorbs 3 kW from a 240 V d.c. supply.

$$P = \frac{U^2}{R}$$
$$3000 = \frac{240^2}{R}$$

$$\therefore \quad \frac{1}{3000} = \frac{R}{240^2}$$

$$\therefore \quad R = \frac{240 \times 240}{3000} = 19.2\,\Omega$$

EXAMPLE 3 Determine the voltage which must be applied to a $9.8\,\Omega$ resistor to produce 500 W of power.

$$P = \frac{U^2}{R}$$

$$500 = \frac{U^2}{9.8}$$

$$\therefore \quad U^2 = 9.8 \times 500$$

$$= 4900$$

$$\therefore \quad U = \sqrt{4900}$$

$$= 70\text{V}$$

EXERCISE 10

1. A contactor coil has resistance of $800\,\Omega$. Calculate the power absorbed by this coil from a 240 V d.c. supply.

2. A piece of equipment creates a voltage drop of 180 V and the power absorbed by it is 240 W. Determine its resistance.

3. Calculate the resistance of a 36 W, 12 V car headlamp bulb.

4. Determine the voltage to be applied to a $6\,\Omega$ resistor to produce 2400 W of power.

5. Complete the following table:

Power (W)		100	60	125		1800		36
Voltage (V)	80	240	250		240	220	3.5	
Resistance (Ω)	50			20	75		0.29	4

6. Calculate the maximum voltage which may be applied to a $45\,\Omega$ resistor rated at 5 W.

7. Determine the power absorbed by the field coils of a 460 V d.c. motor. The resistance of the coils is $380\,\Omega$.

8. Determine the resistance of a 230 V, 1 kW heater.

9. The voltage drop in a cable of resistance $0.072\,\Omega$ is 3.5 V. Calculate the power wasted in the cable.

10. Determine the resistance of a 110 V, 75 W lamp.

11. The following items of equipment are designed for use on a 240 V supply. Calculate the resistance of each item.

    (a) 2 kW radiator   (f) 7 kW cooker

    (b) 3 kW immersion heater   (g) 100 W lamp

    (c) 3.5 kW washing machine   (h) 1500 W kettle

    (d) 450 W toaster   (i) 750 W water heater

    (e) 60 W lamp   (j) 4 kW immersion heater

12. Calculate the voltage drop in a resistor of 12.5 Ω when it is absorbing 500 W.

13. The power dissipated in a 57 Ω resistor is 1000 W. Determine the current.

14. Two lamps are connected in series to a 200 V supply. The lamps are rated at 150 W, 240 V and 60 W, 240 V. Calculate

    (a) the current taken from the supply   (b) the total power.

15. Two 1000 W, 240 V heater elements are connected to a 240 V d.c. supply (a) in series (b) in parallel. Calculate

    (i) the combined resistance in each case,

    (ii) the power absorbed in each case.

16. Cables of resistance 0.35 Ω and 0.082 Ω are connected in parallel and they share a load of 100 A. Determine the current and power loss in each.

17. The element of an immersion heater has a total resistance of 76.8 Ω and is centre-tapped. Calculate the power absorbed from a 240 V supply when the element sections are (a) in series (b) in parallel.

18. Complete the following table and then plot a graph of power (*vertically*) against current (*horizontally*). Try to make the axes of the graph of equal length, and join the points with a smooth curve.

Power (W)		250	360	400	600	
Current (A)	0.8	2.5		3.15		4.9
Resistance (Ω)	40		40		40	40

From the graph, state

    (a) what power would be dissipated in a 40 Ω resistor by a current of 3.7 A,

    (b) how much current is flowing when the power is 770 W?

19. Complete the following table and plot a graph of power against voltage. Join the points with a smooth curve.

Power (W)		2000		750	420	180
Voltage (V)	240	200	180	120		
Resistance (Ω)	19.2		19.5		19.1	19.2

(a) Read off the graph the voltage when the power is 1500 W.

(b) Extend the graph carefully and find from it the power when the voltage is 250 V.

20. The voltage applied to the field circuit of a motor can be varied from 250 V down to 180 V by means of a shunt field regulator. The resistance of the field coils is 360 Ω. Plot a graph showing the relationship between the power and the applied voltage.

21. A cable of resistance 0.07 Ω carries a current which varies between 0 and 90 A. Plot a graph showing the power loss in the cable against the load current.

22. A current of 4.8 A flows in a resistor of 10.5 Ω. The power absorbed is
    (a) 529.2 W   (b) 24 192 W   (c) 2420 W   (d) 242 W

23. The power developed in a resistor of 24 Ω is 225 W. The current flowing is
    (a) 9.68 A   (b) 3.06 A   (c) 0.327 A   (d) 30.6 A

24. The resistance of a 110 V, 100 W lamp is
    (a) 1210 Ω   (b) 0.011 Ω   (c) 8.26 Ω   (d) 121 Ω

25. The voltage to be applied to a resistor of 55 Ω in order to develop 50 watts of power is
    (a) 0.95 V   (b) 166 V   (c) 52.4 V   (d) 1.05 V

Electrical charge, energy and tariffs

ELECTRICAL CHARGE

The charge or quantity of electricity conveyed by a current in a given interval of time is

$$Q = I \times t \text{ coulombs}$$

where I is the current in amperes and t is the time in seconds.

EXAMPLE 1   Find the charge conveyed by a current of 2.5 A flowing for 15 s.

$$Q = I \times t$$
$$= 2.5 \times 15$$
$$= 37.5 \text{ C}$$

A larger, non-SI unit, the ampere hour (A h), is generally used in connection with the storage capacity and the charging of batteries.

EXAMPLE 2   A 120 A h battery is to be charged fully over a period of 10 hours. Calculate the average charging current.

$$Q = I \times t$$
$$120 \text{ A h} = I \times 10 \text{ h}$$
$$I = \frac{10 \cancel{h}}{120 \text{ A} \cancel{h}} \quad \text{(note the cancelling of the units not required in the answer)}$$
$$= 12 \text{ A}$$

ELECTRICAL ENERGY

The energy supplied to an electric circuit in a given interval of time is

$$W = P \times t \text{ joules}$$

where P is the power in watts and t is the time in seconds. (P may be calculated in several ways as previously shown.)

The following are useful forms:

$$W = UI \times t \text{ joules}$$

$$W = I^2 R \times t \text{ joules}$$

$$W = \frac{U^2}{R} \times t \text{ joules}$$

where U is in volts, R is in ohms, and I is in amperes.

EXAMPLE Calculate the energy dissipated in a $10\,\Omega$ resistor when a current of 1.5 A flows for 25 s.

$$W = I^2 R \times t$$
$$= 1.5 \times 1.5 \times 10 \times 25$$
$$= 562.5\,\text{J}$$

A larger unit, the kilowatt hour (kW h), is used as the basis of the cost of electrical energy supplied to a consumer. It is often referred to simply as a 'unit'.

ELECTRICITY TARIFFS

Electricity suppliers publish various tariffs for the supply of electrical energy to consumers. These tariffs reflect the cost of producing the energy at the power station, distributing the energy, via underground and overhead cables, and providing services to the consumer's premises including metering, etc.

In addition to the above basic charges for supply (standing) costs and energy consumed costs, the government levies value added tax. At present this stands at 5% of total charges.

Electricity suppliers render accounts monthly, quarterly or in the case of commercial and industrial consumers at agreed intervals.

EXAMPLE I Calculate the basic cost of operating a 3 kW heater for 15 hours if energy costs 7.4p per unit (kW h).

$$\text{Number of kilowatt hours} = 3\,\text{kW} \times 15\,\text{hours}$$
$$= 45\,\text{kW h}$$
$$\text{Cost} = 45 \times 7.4\text{p}$$
$$= 333\text{p}$$
$$= £3.33$$

If 5% VAT is chargeable the actual cost will be

$$£3.33 \times 1.05 = £3.50$$

Thus the inclusive cost per unit (basic plus VAT) is

$$\frac{350}{45} = 7.78\text{p}$$

EXAMPLE 2 Alternative tariffs are available to a domestic consumer as follows:

(a) a standing charge of 11.09p per day plus 7.8p per unit used,

(b) a standing charge of 8.03p per day plus 4.1p per unit used between midnight and 8.00 am plus 8.9p for each unit used at any other time of the day.

In a particular quarter (90 days) the consumer used 4140 units, of which 1500 units are used between midnight and 8.00 am. Calculate the cost of the electrical energy supplied under each tariff. VAT is chargeable at 5%.

(a) Standing charge   90 × 11.09p   = £9.98

    Energy cost   4140 units at 7.8p = £322.92

                 Total basic cost   = £332.90

                 adding 5% VAT   = £332.90 × 1.05

                 Total charge   = £349.545   or   £349.55

(b) Standing charge   90 × 8.03   = £7.23

    1500 units at 4.1p   = £61.50

    (4140 − 1500) units at 8.9p per unit:

                 2640 × 8.9   = £234.96

                 Total basic cost   = £296.46

                 adding 5% VAT   = £296.46 × 1.05

                               = £311.283   or   £311.28

EXERCISE 11

Add VAT at 5% where electrical energy is purchased from an electricity supplier. Omit VAT where unit charge is inclusive.

1. A current of 1.75 A flows for 35 s. Calculate the quantity of electricity involved.

2. For how long must a current of 0.6 A flow for a charge of 120 C to accumulate?

3. Complete the following table:

Current (A)	0.025		0.55			8.5
Time (s)	125	7.5			360	
Charge (C)		115	1050	4200		540

4. A current of 15 A flows for 1 hour. Calculate the charge (a) in coulombs (b) in ampere hours.

5. A battery requires 250 A h of charge over a period of 8 hours. Calculate the average charging current.

6. A 6 Ω resistor is connected to a 12 V supply for 10 minutes. Calculate the energy dissipated.

7. A current of 1.5 A flows in a 15 Ω resistor for 6 minutes. Calculate the energy dissipated.

8. A current of 4.5 A flows from a 240 V supply for 8 hours. Calculate (a) the quantity of electricity in coulombs and ampere hours, (b) the energy supplied in joules and kilowatt hours.

9. Find the cost of operating a 2 kW radiator 6 hours per day for seven days with energy at 7.3p per unit (inclusive).

10. A 3 kW immersion heater is switched on for three hours in every four. If the electrical energy is 7.3p per unit (inclusive), how much does it cost per day of 24 hours?

11. Calculate the running cost per hour for a 7 kW cooker, assuming that only 0.6 of its loading is in use. Electrical energy is 7.7p per unit (inclusive).

12. Find the cost per hour of operating each of the following if energy is 7.6p per unit (inclusive):
  (a) a 3 kW immersion heater,
  (b) a 150 W lamp,
  (c) a 450 W iron,
  (d) a vacuum cleaner taking 350 W,
  (e) a hi-fi set taking 70 W.

13. Electrical energy costs 6.8p per unit (inclusive). For how long may each of the following items of equipment be operated for 20p?
  (a) a 60 W lamp,
  (b) a 2 kW radiator,
  (c) a television set taking 200 W,
  (d) a refrigerator taking 270 W,
  (e) an electric clock taking 25 W.

14. In a house, the following items of equipment are switched on together for 6 hours each day: two 100 W lamps, three 60 W lamps, one 2 kW radiator and one 3 kW immersion heater.

    The quarterly charge for the energy is based on a tariff in which there is a standing charge of £8.90 plus 7.3p per unit used. Find, to the nearest penny, the electricity bill for a quarter of 90 days.

15. Tubular heaters having a total loading of 6 kW are installed in a glasshouse. The control device switches the heaters on for one hour and then off for half an hour and it operates continuously. Find the cost of electrical energy at 7.9p per unit (inclusive) used in a week of 7 days.

16. A floodlit sign is provided with four 750 W lamps and they are switched on at 19.00 hours and off at 23.00 hours each evening. The charge for the energy is as follows:

    each of the first 20 units per 100 W installed   15.58p

    each additional unit   6.8p

    Determine the cost of illuminating the sign for a period of 120 days.

17. Determine the approximate cost of the electricity used in a private house during a quarter of 90 days where the charge is made as follows: a fixed charge of £9.66 plus 7.3p per unit. The apparatus in use is

    two 100 W lamps in use 5 hours per day,

    three 60 W lamps in use 2 hours per day,

    one 3 kW immersion heater in use 8 hours per week,

    one 3 kW washing machine in use 5 hours per week.

18. Find the cost of running a motor for a 10 hour day if it takes 17.5 A from a 400 V supply and energy is 8.3p per unit (inclusive).

19. An electric furnace requires a current of 14.5 A from a 230 V supply for $1\frac{1}{2}$ hours in order to heat a certain quantity of steel. How much does the process cost if the electrical energy is 10.3p per unit (inclusive)?

20. The energy used in a certain electric welding process costs 6.8p per unit (inclusive). The process requires a current of 120 A at 50 V for 15 seconds. Find the cost of 1000 such operations.

21. A diesel generator set delivered a current of 80 A at 230 V for 100 hours. The cost of running the set during this time was £475. How much has each unit of electricity cost?

22. A pump raises 0.15 m^3 of water per minute from a well. The motor driving the pump takes a current of 4.2 A from a 240 V supply. The

electrical energy costs 7.3p per unit. How much will it cost to pump 100 m³ of water?

23. A circular saw takes 90 seconds to cut through a 6 m length of timber. It is driven by an electric motor which during this time takes a current of 15 A from a 230 V supply. If the electrical energy costs 7.6p per unit (inclusive), find the cost of making 250 such cuts.

24. The following apparatus is in use in a workshop for 8 hours each day. Electrical energy is 10.3p per unit (inclusive). Find to the nearest penny the electricity bill for a five-day week.

Apparatus	Loading
3 motors	4 A each at 240 V
6 heating units	2 kW each
I boiler	6 kW
8 fluorescent lights	80 W each

25. To heat a certain building would require electric heaters taking 125 A from a 230 V supply for 12 hours. Alternatively, a coal-fired boiler could be used, in which case the coal would cost £9.20 and a stoker would have to be employed for 5 hours at £4.50 per hour. State which method of heating would be the cheaper and show calculations to justify your choice. Take the cost of electrical energy as 7.3p per unit (inclusive).

26. A domestic consumer has the following connected loads:

space heating, 20 kW in use daily between midnight and 0630 hours;

one 3 kW immersion heater in use 8 hours per week;

six 100 W lamps in use 28 hours per week.

The tariff is as follows:

fixed charge per quarter   £9.66

unit charge:

   each unit supplied for space heating between
      midnight and 0800 hours   2.79p

   each unit supplied for other purposes   7.34p

Calculate the electricity account for a quarter containing thirteen weeks.

27. A charge of 1800 C is conveyed by a current in 25 seconds. The current is

(a) 72 A   (b) 45 000 A   (c) 0.014 A   (d) 4320 A

28. A current of 4.5 A flows in a 3.5 Ω resistor for 5.5 minutes. The energy dissipated is

(a) 476 J   (b) 390 J   (c) 23 389 J   (d) 18 191 J

29. A 2.5 kW heater is in use for 7.5 hours daily and energy costs 7.1p per unit. The total cost of running the heater over a month of 28 days is
(a) £372.75   (b) £133.31   (c) £37.28   (d) £1.33

30. An installation has the following connected loads continuously in use: a 3 kW heater, 2 kW of lighting, motors taking 10 kW.

Over a 40 h period the total cost of energy was £35.40. Each unit cost
(a) 8.85p   (b) 5.9p   (c) 0.885p   (d) 0.59p

Percentages and efficiency

EXAMPLE 1 The voltage at the terminals of a motor is 435 V and the mains voltage is 450 V. Determine the percentage voltage drop.

$$\text{Percentage voltage drop} = \frac{450 - 435}{450} \times 100$$

$$= \frac{15}{450} \times 100$$

$$= 3.33\%$$

EXAMPLE 2 In accordance with BS 7671 requirements, the maximum voltage drop allowed in a normal circuit is 4% of the supply voltage at the origin of the installation. Calculate the maximum voltage drop allowed in a circuit fed at 230 V.

$$\text{Maximum voltage drop} = 4\% \text{ of } 230\,\text{V}$$

$$= \frac{4}{100} \times 230$$

$$= 9.2\,\text{V}$$

EXAMPLE 3 The output of a generator is 1500 W and the input is equivalent to 1900 W. Calculate its percentage efficiency.

$$\text{Efficiency} = \frac{\text{output}}{\text{input}}$$

$$= \frac{1500}{1900}$$

$$= 0.789$$

$$\text{Percentage efficiency} = 0.789 \times 100$$

$$= 78.9\%$$

EXERCISE 12

1. To comply with BS 7671 requirements, the voltage drop in a circuit must not exceed 4% of the supply voltage. Determine the maximum voltage drop permissible if the supply voltage is (a) 250 V, (b) 415 V, (c) 110 V.

2. It is recommended that in a domestic installation an allowance should be made for 66% of the lighting equipment being in use at the same time. If in a certain house the total lighting load is 530 W, estimate how much of this is likely to be in use at any time.

3. In the case of a circuit supplying sensitive computer equipment a maximum voltage variation of 2.5% is specified. Calculate (a) the lowest and (b) the highest voltage permitted if the rated voltage of the equipment is 230 V.

4. An automatic voltage regulator is guaranteed to keep the voltage generated by a 230 V alternator constant to within ±2%. Calculate the highest and lowest values of the voltage.

5. 3.5 kW of power is supplied to a certain installation, but $1\frac{1}{2}\%$ of the power is wasted in the supply cables. How much power is actually used in the installation?

6. The voltage at the terminals of a motor is 223 V, whereas the mains voltage is 230 V. Calculate the percentage voltage drop.

7. A load current of 56 A is supplied from a 220 V main through cables which have a total resistance of 0.027 Ω. Calculate the percentage voltage drop.

8. The voltage at the terminals of a certain heater is 235 V and the current flowing is 15.5 A. The supply point is some distance away and the voltage there is 242 V. Calculate
   (a) the power used by the heater,
   (b) the power wasted in the cables,
   (c) the percentage of power supplied which is wasted in the cables.

9. A heater takes a current of 12.5 A from a 230 V supply and there is a 1.5% drop in voltage in the supply cables. Determine the resistance of the cables.

10. A voltmeter which can indicate up to 250 V is only guaranteed to be accurate to within ±1%. Determine the highest and lowest voltages that could be present when the reading is 240 V.

11. The overload release on a motor starter is set to trip at 175% full-load current and it is found that it actually operates when the current through it is 5.6 A. Calculate the normal full-load current of the motor.

12. The power output from a generator is 2600 W and the power required to drive it is equivalent to 3500 W. Determine the percentage efficiency of the generator.

13. The power output of a motor is 250 W and the input is 1.5 A at 240 V. Calculate the percentage efficiency of the motor.

14. A generator supplies 20 lamps, each taking 0.45 A at 230 V. The power input to the machine is 2850 W. Calculate the percentage efficiency of the generator.

15. A motor takes a current of 12.5 A from a 460 V supply and its efficiency is 76%. Determine the power output of the motor in kilowatts.

16. Calculate the current taken by a motor whose output power is 2.5 kW and efficiency 72% when it is working from a 240 V supply.

17. A special-purpose heater having a resistance of 3 Ω is supplied from a battery of e.m.f. 6 V and internal resistance 0.48 Ω. Calculate (a) the power dissipated in the heater, (b) the percentage of the total power wasted in the internal resistance.

18. A 3 kW immersion heater (rated at 230 V) is in use in a certain installation but, due to voltage drop, the voltage at its terminals is only 218.5 V. Calculate:
   (a) the percentage voltage drop,
   (b) the actual power consumed by the heater,
   (c) the percentage reduction in the power of the heater caused by the voltage drop.

19. The load current required by a certain installation is 160 A and it is supplied from the 230 V mains through cables which have a total resistance of 0.022 Ω. Determine
   (a) the percentage voltage drop,
   (b) the percentage of the total power supplied which is wasted in the cables.

20. A d.c. generator is rated at 2.5 kW but it will operate with a 15% overload. What is the maximum number of 75 W lamps it can supply?

21. A current of 45 A flows in cables which have a total resistance of 0.293 Ω. The supply voltage is 240 V. The percentage voltage drop is
   (a) 0.7%   **(b)** 64%   **(c)** 5.5%   **(d)** 0.55%

22. The voltage at the terminals of a heater is 234.5 V and the supply voltage is 240 V. The percentage voltage drop is
   (a) 2.3%   **(b)** 0.44%   **(c)** 2.35%   **(d)** 23%

Areas, space factor and volumes

AREAS AND PERIMETERS

Rectangle
Perimeter $P = 2l + 2b$
Area $A = l \times b$

Triangle
Area $A = \frac{1}{2}b \times h$

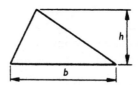

Circle

Circumference $C = \pi d$

Area $A = \dfrac{\pi d^2}{4}$

EXAMPLE 1   Calculate the cross-sectional area of trunking whose dimensions are 50 mm by 38 mm.

$A = l \times b$

    $= 50 \, \text{mm} \times 38 \, \text{mm}$

    $= 1900 \, \text{mm}^2$

EXAMPLE 2   The rectangular bottom of a tank is 0.5 m by 0.75 m. Calculate (a) its area, (b) the length of the weld forming its perimeter.

(a)   Area $A = l \times b$

           $= 0.5 \, \text{m} \times 0.75 \, \text{m}$

           $= 0.375 \, \text{m}^2$

(b)   Perimeter $P = 2l + 2b$

            $= (2 \times 0.75) \, \text{m} + (2 \times 0.5) \, \text{m}$

            $= 2.5 \, \text{m}$

EXAMPLE 3   The triangular gable end of a house is 8 m wide and 3.5 m high. Calculate its area.

$A = \tfrac{1}{2} b \times h$

   $= \tfrac{1}{2} \times 8 \, \text{m} \times 3.5 \, \text{m}$

   $= 14 \, \text{m}^2$

EXAMPLE 4   Calculate the nominal cross-sectional area of 7/1.70 pvc-insulated cable if the nominal diameter of this cable is 7.3 mm.

$$\text{Nominal cross-section area } A = \frac{\pi d^2}{4}$$

$$= \frac{\pi \times 7.3^2}{4}$$

$$= 41.85\,\text{mm}^2$$

EXAMPLE 5   A copper cylinder is 0.5 m in diameter. Calculate
(a) the area of the bottom,
(b) the length of lagging material required to surround it.

(a) $A = \dfrac{\pi d^2}{4}$

$\quad\quad = \dfrac{\pi \times 0.5 \times 0.5}{4}$

$\quad\quad = 0.196\,\text{m}^2$

(b) $C = \pi \times d$

$\quad\quad = \pi \times 0.5$

$\quad\quad = 1.57\,\text{m}$

EXAMPLE 6   Convert $3500\,\text{mm}^2$ to m^2.

$$3500\,\text{mm}^2 = 3500\,\cancel{\text{mm}} \times \cancel{\text{mm}} \times \left[\frac{1\,\text{m}}{10^3\,\cancel{\text{mm}}}\right] \times \left[\frac{1\,\text{m}}{10^3\,\cancel{\text{mm}}}\right]$$

$$= \frac{3500}{10^6}\,\text{m}^2 = 3.5 \times 10^{-3}\,\text{m}^2$$

SPACE FACTOR

EXAMPLE I   Space factor is the ratio (expressed as a percentage) of the sum of the effective overall cross-sectional areas of cables (including insulation and any sheath) to the internal cross-sectional area of the conduit, duct, trunking, or channel in which they are installed.

A 20 mm diameter conduit contains twelve $1.0\,\text{mm}^2$ single-core pvc cables. Determine the space factor.

From Table A, the nominal overall diameter of a 1.0 mm^2 cable is 2.9 mm.

$$\therefore \quad \text{cross-sectional area of one cable, } A = \frac{\pi d^2}{4}$$

$$= \frac{\pi \times 2.9^2}{4}$$

$$\therefore \quad \text{total area of 12 cables} = 12 \times \frac{\pi \times 2.9^2}{4} \, \text{mm}^2$$

$$= 25.23\pi \, \text{mm}^2 \qquad \text{(i)}$$

(Do not evaluate π in your calculation at this stage.)

The inside diameter of 20 mm steel conduit is about 17 mm,

$$\therefore \quad \begin{array}{l} \text{inside cross-sectional} \\ \text{area of the conduit} \end{array} = \frac{\pi \times 17^2}{4}$$

$$= 72.25\pi \, \text{mm}^2 \qquad \text{(ii)}$$

$$\text{space factor} = \text{(i)} \div \text{(ii)} = \frac{25.23\pi \, \text{mm}^2}{72.25\pi \, \text{mm}^2} \qquad \text{(iii)}$$

$$= 0.3492$$

or $\quad 0.3492 \times 100\% = 34.9\%, \quad$ say 35%

Note the π cancels in equation (iii), which is the reason for not working it out earlier.

EXAMPLE 2 Determine the number of 35 mm^2 single-core pvc cables that may be installed in a 50 mm $\times$ 50 mm trunking, allowing a space factor of 45%.

$$\text{Cross-sectional area of the trunking} = 50 \, \text{mm} \times 50 \, \text{mm}$$

$$= 2500 \, \text{mm}^2$$

The total cross-sectional area of cables permissible within the 45% space factor is

$$45\% \text{ of } 2500 \, \text{mm}^2 = \frac{45}{100} \times 2500 = 1125 \, \text{mm}^2 \qquad \text{(i)}$$

From Table A, the nominal overall diameter of the 35 mm^2 cable is 10.3 mm,

$\therefore$ cross-sectional area of one cable $= \dfrac{\pi d^2}{4}$

$$= \frac{3.142 \times 10.3^2}{4}$$

$$= 83.33 \text{ mm}^2 \qquad \text{(ii)}$$

The permitted number of cables is (i) $\div$ (ii)

$$= \frac{1125}{83.33}$$

$$= 13.5 \quad \text{or} \quad 13$$

Table A

Details of single-core pvc-insulated cables

Cable size		
Nominal conductor size (mm^2)	*Number and diameter of wires* (no./mm)	*Nominal overall diameter* (mm)
1.0	1/1.13	2.9
1.5	1/1.38	3.1
2.5	1/1.78	3.5
2.5	7/0.67	3.8
4	7/0.85	4.3
6	7/1.04	4.9
10	7/1.35	6.2
16	7/1.70	7.3
25	7/2.14	9.0
35	19/1.53	10.3
50	19/1.78	12.0

Table B

Dimensions of trunking (mm × mm)

50 × 37.5
50 × 50
75 × 25
75 × 37.5
75 × 50
75 × 75
100 × 25
100 × 37.5
100 × 50
100 × 75
100 × 100

EXAMPLE 3   The following cables are to form part of an installation: ten $1.0\,\text{mm}^2$; six $2.5\,\text{mm}^2$; three $50\,\text{mm}^2$; one $25\,\text{mm}^2$. It is proposed to instal them all in a single run of trunking. Determine the size of trunking required.

The total cross-sectional area of cable is calculated as follows, using values from Table A.

$$\text{nominal diameter of } 1.0\,\text{mm}^2 \text{ cable} = 2.9\,\text{mm}$$

$$\therefore \quad \text{cross-sectional area of } 1.0\,\text{mm}^2\text{cable} = 6.606\,\text{mm}^2$$

(see example 4
on page 62)

$$\therefore \quad \text{c.s.a. of ten such cables} = 10 \times 6.606$$

$$= 66.06\,\text{mm}^2$$

$$\text{nominal diameter of } 2.5\,\text{mm}^2 \text{ cable} = 3.8\,\text{mm}$$

$$\therefore \quad \text{c.s.a. of } 2.5\,\text{mm}^2 \text{ cable} = 11.34\,\text{mm}^2$$

$$\therefore \quad \text{c.s.a. of six such cables} = 68.04\,\text{mm}^2$$

$$\text{nominal diameter of } 50\,\text{mm}^2 \text{ cable} = 12\,\text{mm}$$

$$\therefore \quad \text{c.s.a. of } 50\,\text{mm}^2 \text{ cable} = 113.11\,\text{mm}^2$$

$$\therefore \quad \text{c.s.a. of three such cables} = 339.33\,\text{mm}^2$$

$$\text{nominal diameter of } 25\,\text{mm}^2 \text{ cable} = 9\,\text{mm}$$

$$\therefore \quad \text{c.s.a. of one } 25\,\text{mm}^2 \text{ cable} = 63.63\,\text{mm}^2$$

$$\text{Total cross-sectional area of cable} = 66.06 + 68.04$$

$$+\, 339.33 + 63.63$$

$$= 537\,\text{mm}^2$$

Allowing for the required 45% space factor,

$$\text{space required} = 537 \times \frac{100}{45}$$

$$= 1193\,\text{mm}^2$$

and $50\,\text{mm} \times 37.5\,\text{mm}$ or $75 \times 25\,\text{mm}$ trunking (with cross-sectional area $1875\,\text{mm}^2$) is suitable.

BS 7671, the IEE On-Site Guide and many manufacturers' catalogues contain tables which simplify the selection of conduit and trunking to contain single-core pvc-insulated cables. The tables list cable sizes, conduit diameters and trunking dimensions and allocate a factor for each. Thus the tables give rapid guidance to the size of enclosure required to accommodate multiple cable runs.

The conduit tables also give guidance to the cable capacity for various lengths of run between draw-in boxes and for conduit with bends in their run.

EXAMPLE 1   A conduit is required to contain ten single-core $1.0\,\text{mm}^2$ pvc-insulated cables. The length of conduit between the control switches and an electric indicator-lamp box is 5 m, and the conduit run has two right-angle bends. Select a suitable size of conduit.

From table 5C (IEE On-Site Guide), the cable factor for ten $1.0\,\text{mm}^2$ cables $= 10 \times 16 = 160$.

From table 5D (IEE On-Site Guide), for a 5 m run with two bends select 20 mm conduit with a conduit factor of 196.

EXAMPLE 2   Steel conduit is required to contain the following stranded-conductor single-core pvc-insulated cables for a machine circuit:

(a) three $6\,mm^2$ cables between a steel cable trunking and the machine control box,

(b) in addition to the $6\,mm^2$ cables between the control box and the machine there are three $2.5\,mm^2$ cables and six $1.5\,mm^2$ cables.

The conduit from the cable trunking to the machine control box is 1.5 m long with one bend, and the steel conduit between the control box and the machine is 3.5 m with two bends.

Using appropriate cable and conduit tables, select suitable conduit sizes for:

(i) trunking to control box and

(ii) control box to machine.

(i) From table 5A (IEE On-Site Guide) cable factor for $6\,mm^2$ cable is 58, thus cable factor for three
$6\,mm^2$ cables $= 58 \times 3 = 174$.
From table 5D (IEE On-Site Guide) for a conduit run of 2.5 m with one bend select 20 mm conduit with a factor of 278.

(ii) From table 5C (IEE On-Site Guide) cable factor for $6\,mm^2$ cable is 58, thus cable factor for three
$6\,mm^2$ cables $= 58 \times 3 = 174$. Again from table 5C cable factor for $2.5\,mm^2$ cable is 30, thus cable factor for three
$2.5\,mm^2$ cables $= 30 \times 3 = 90$ and from table 5C cable factor for $1.5\,mm^2$ cable is 22, thus cable factor for six
$1.5\,mm^2$ cables $= 22 \times 6 = 132$.

Thus total cable factors $= 174 + 90 + 132 = 396$.

From table 5D (IEE On-Site Guide) for a 3.5 m conduit run with two bends, select 25 mm conduit with a factor of 404.

EXERCISE 13

1. The floor of room is in the form of a rectangle 3 m by 3.5 m. Calculate its area.

2. A rectangular electrode for a liquid resistor is to have area $0.07\,mm^2$. If it is 0.5 m long, how wide must it be?

3. Complete the table below, which refers to various rectangles:

Length (m)	6		12	8	
Breadth (m)	2	2			
Perimeter (m)		10		24	32
Area (m²)			84		48

4. The triangular portion of the gable end of a building is 6 m wide and 3.5 m high. Calculate its area.

5. The end wall of a building is in the form of a square with a triangle on top. The building is 4 m wide and 5.5 high to the top of the triangle. Calculate the total area of the end wall.

6. Complete the table below, which refers to various triangles:

Base (m)	0.5	4	1.5		0.3
Height (m)	0.25		2.2	3.2	0.12
Area (m²)		9		18	

7. Complete the following table:

Area (m²)	0.015			0.000 29	0.0016
Area (mm²)		250	7500		

8. Complete the table below, which refers to various circles:

Diameter	0.5 m				4 mm
Circumference		1.0 m			
Area			0.5 m²	6 mm²	

9. A fume extract duct is to be fabricated on site from aluminium sheet. Its dimensions are to be 175 mm diameter and 575 mm length. An allowance of 25 mm should be left for a rivetted joint along its length. Establish the area of metal required and the approximate number of rivets required, assuming rivets at approximately 70 mm spacing.

10. A square ventilation duct is to be fabricated on site from steel sheet. To avoid difficulty in bending, the corners are to be formed by 37.5 mm × 37.5 mm steel angle and 'pop' rivetting. Its dimensions are to be 259 mm × 220 mm × 660 mm length. Establish the area of sheet steel, length of steel angle and the approximate number of rivets required, assuming rivets at 60 mm spacing.

11. A coil of wire contains 25 turns and is 0.25 m in diameter. Calculate the length of wire in the coil.

12. Complete the table below, which refers to circular conductors:

Number and diameter of wires (mm)	1/1.13		7/0.85		
Nominal cross-sectional area of conductor (mm²)		2.5		10	25

13. Complete the table below, which refers to circular cables:

Nominal overall diameter of cable (mm)	2.9	3.8	6.2	7.3	12.0
Nominal overall cross-sectional area (mm^2)					

14. Calculate the cross-sectional areas of the bores of the following heavy-gauge steel conduits, assuming that the wall thickness is 1.5 mm:

(a) 16 mm (b) 25 mm (c) 32 mm

15. Complete the following table, using a space factor of 45% in each case:

	Permitted number of pvc cables in trunking of size (mm)		
Cable size	50 × 37.5	75 × 50	75 × 75
16 mm^2			
25 mm^2			
50 mm^2			

16. The following pvc cables are to be installed in a single run of trunking: twelve 16 mm^2, six 35 mm^2, twenty-four 2.5 mm^2, and eight 1.5 mm^2.

Determine the size of trunking required, assuming a space factor of 45%.

17. Determine the size of square steel trunking required to contain the following pvc cables: fifteen 50 mm^2, nine 25 mm^2, eighteen 10 mm^2. Take the space factor for ducts as 35%.

18. The nominal diameter of a cable is 6.2 mm. Its cross-sectional area is

(a) 120.8 mm^2 (b) 19.5 mm^2 (c) 30.2 mm^2 (d) 61.2 mm^2

19. Allowing a space factor of 45%, the number of 50 mm^2 cables that may be installed in a 50 mm × 37.5 mm trunking is

(a) 71 (b) 8 (c) 23 (d) 37

The following cable calculations require the use of data contained in documents based upon BS 7671, e.g. IEE On-Site Guide, etc. In each case assume that the stated circuit design calculations and environmental considerations have been carried out to determine the necessary cable current ratings and type of wiring system.

20. A steel cable trunking is to be installed to carry eighteen $1.5\,\text{mm}^2$ single-core pvc-insulated cables to feed nine floodlighting luminaires; a single $4\,\text{mm}^2$ protective conductor is to be included in the trunking. Establish the minimum size of trunking required.

21. $50\,\text{mm} \times 38\,\text{mm}$ pvc trunking is installed along a factory wall to contain low-current control cables. At present there are 25 pairs of single-core $1.5\,\text{mm}^2$ pvc-insulated cables installed. How many additional pairs of similar $1.5\,\text{mm}^2$ control cables may be installed in the trunking?

22. A pvc conduit is to be installed to contain six $4\,\text{mm}^2$ single-core pvc cables and one $2.5\,\text{mm}^2$ stranded single-core pvc protective conductor. The total length of run will be $16\,\text{m}$ and it is anticipated that four right-angle bends will be required in the conduit run. Determine the minimum conduit size and state any special consideration.

23. An electric furnace requires the following wiring:

(i) three $6\,\text{mm}^2$ stranded single-core pvc cables,

(ii) four $2.5\,\text{mm}^2$ stranded single-core pvc cables,

(iii) four $1.5\,\text{mm}^2$ stranded single-core pvc cables.

There is a choice between new steel conduit and using existing $50\,\text{mm} \times 38\,\text{mm}$ steel trunking which already contains six $25\,\text{mm}^2$ single-core pvc cables and four $10\,\text{mm}^2$ single-core pvc cables. Two right-angle bends will exist in the $18\,\text{m}$ run.

(a) Determine the minimum size of conduit to be used, and

(b) state whether the new cables could be included within the existing trunking, and if they could be, what considerations must be given before their inclusion.

24. Select two alternative sizes of steel trunking which may be used to accommodate the following.

(i) ten $16\,\text{mm}^2$ single-core pvc-insulated cables,

(ii) twelve $6\,\text{mm}^2$ single-core pvc-insulated cables,

(iii) sixteen $1.5\,\text{mm}^2$ single-core pvc-insulated cables,

(iv) three multicore pvc-insulated signal cables, assuming a cable factor of 130.

An extension to the trunking contains ten of the $16\,\text{mm}^2$ cables and 8 of the $1.5\,\text{mm}^2$ cables.

Establish the minimum size of conduit, assuming a $5\,\text{m}$ run with no bends. How may the conduit size selected affect the choice of trunking dimensions (assume that the two sizes of trunking cost the same).

VOLUMES

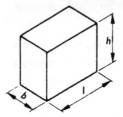

Rectangular object   Cylindrical object

Volume $V = l \times b \times h$ Volume $V = \dfrac{\pi d^2}{4} \times h$

The SI unit of volume is the cubic metre (m^3) with smaller units dm^3, cm^3 and mm^3. Liquids are measured in litres (1 litre = 1 dm^3) with smaller units ml, cl.

EXAMPLE 1   A tank is 0.35 m by 0.3 m by 0.6 m. Find the mass of oil it contains when completely full. The relative density of oil is 0.9.

$$\text{Volume of oil, } V = l \times b \times h$$
$$= 0.35\,\text{m} \times 0.3\,\text{m} \times 0.6\,\text{m}$$
$$= 0.063\,\text{m}^3 \ (63\,\text{litres})$$
$$\text{Mass of } 1\,\text{m}^3 \text{ of water} = 10^3\,\text{kg}$$
$$\text{mass of } 0.063\,\text{m}^3 \text{ of water} = 0.063 \times 10^3\,\text{kg}$$
$$\therefore \quad \text{mass of } 0.063\,\text{m}^3 \text{ of oil} = 0.063 \times 10^3 \times 0.9\,\text{kg}$$
$$= 56.7\,\text{kg}$$

EXAMPLE 2   Calculate the capacity of a cylindrical water tank 0.3 m in diameter and 0.6 m high.

$$V = \dfrac{\pi d^2}{4} \times h$$
$$= \dfrac{\pi \times 0.3^2}{4} \times 0.6$$
$$= 0.042\,\text{m}^3 \quad \text{or} \quad 42\,\text{litres}$$

EXAMPLE 3 Convert $37\,500\,\text{mm}^3$ to m^3.

$$37\,500\,\text{mm}^3 = 37\,500\,\text{mm} \times \text{mm} \times \text{mm}$$

$$37\,500 \times \left[\frac{1\,\text{m}}{10^3\,\text{mm}}\right] \times \left[\frac{1\,\text{m}}{10^3\,\text{mm}}\right] \times \left[\frac{1\,\text{m}}{10^3\,\text{mm}}\right]$$

$$= \frac{37\,500}{10^9}\,\text{m}^3$$

$$= 37.5 \times 10^{-6}\,\text{m}^3$$

EXERCISE 14

1. Find the volume of air in a room 5 m by 3.5 m by 2.6 m.

2. Calculate the volume of a cylindrical tank 0.5 m in diameter and 0.75 m long.

3. Find the volume and total surface area of the following enclosed tanks:
   (a) rectangular, 1 m × 0.75 m × 0.5 m;
   (b) cylindrical, 0.4 m in diameter and 0.5 m high.

4. Find the volume and mass of copper in a bar 6 m long and 25 mm by 7.5 mm in cross-section. (Density of copper $= 8.63 \times 10^3\,\text{kg/m}^3$.)

5. Calculate the volume and mass per metre length of a 25 mm diameter round copper bar.

6. The end wall of a building takes the form of a rectangle 15 m wide and 5 m high. It is surmounted by a triangle 3 m high. The building is 25 m long. Determine the volume and mass of air in the room. (Density of air $= 1.28\,\text{kg/m}^3$.)

7. A coil of 6 mm² bare copper wire has mass 5 kg. Calculate the length of wire in the coil.

8. Calculate the mass of 100 m of bare aluminium wire 4 mm² in cross-section. (Density of aluminium $= 2.688 \times 10^3\,\text{kg/m}^3$.)

9. A storage tank has internal dimensions of 526 mm × 630 mm × 1240 mm. Establish (a) its water capacity, allowing for 15% free space; (b) the mass of water it contains.

10. A circular oil storage tank has an external diameter of 526 mm and external length 1360 mm. It is made from 1.5 mm thick metal. Establish its approximate capacity and mass of oil it will contain. Assume oil has a relative density of 0.85.

Resistivity

The resistance of a conductor is

$$R = \frac{\rho \times l}{A} \, \Omega$$

where ρ is the resistivity (Ω m),

l is the conductor length (m)

and A is the cross-sectional area of the conductor (m^2).

EXAMPLE I Determine the resistance of 100 m of 120 mm^2 single-core mineral-insulated cable. (Resistivity of copper $= 1.78 \times 10^{-8} \, \Omega$ m)

$$R = \frac{\rho \times l}{A}$$

$$= \frac{1.78 \times 100}{10^8 \times 120/10^6} \quad \text{(note conversion of } mm^2 \text{ to } m^2 \text{)}$$

$$= \frac{1.78 \times 100 \times 10^6}{10^8 \times 120} = 0.0148 \, \Omega$$

EXAMPLE 2 If, as an alternative, the resistivity of the material is quoted as 17.8 $\mu\Omega$ mm (microhm millimetre), the calculation proceeds as follows:

$$R = \frac{\rho \times l}{A}$$

$$= \frac{17.8 \, \mu\Omega \, \text{mm}}{120 \, \text{mm}^2} \times 100 \, \text{m} \times \left[\frac{1000 \, \text{mm}}{1 \, \text{m}} \right]$$

$$= \frac{17.8 \times 100 \times 1000}{120} \, \mu\Omega$$

$$= 14\,800 \, \mu\Omega$$

$$= 14\,800 \, \mu\Omega \left[\frac{1 \, \Omega}{10^6 \, \mu\Omega} \right]$$

$$= 0.0148 \, \Omega$$

EXAMPLE 3 Calculate the cross-sectional area of an aluminium cable 50 m long which has a resistance of 0.067 Ω. (Resistivity of aluminium $= 2.84 \times 10^{-8}\,\Omega\,\text{m}$.)

$$R = \frac{\rho \times l}{A}$$

$$0.067 = \frac{2.84 \times 50}{10^8 \times A}$$

$$\therefore \quad A = \frac{2.84 \times 50}{10^8 \times 0.067}$$

$$= \frac{21.19}{10^8}\,\text{m}^2$$

$$= \frac{21.19}{10^8} \times 10^6\,\text{mm}^2$$

$$= 21.19\,\text{mm}^2 \quad \text{or} \quad 21.2\,\text{mm}^2$$

EXERCISE 15

In the following exercises, take the resistivity of copper as $1.78 \times 10^{-8}\,\Omega\,m$ and that of aluminium as $2.84 \times 10^{-8}\,\Omega\,m$.

1. Determine the resistance of 100 m of copper cable whose cross-sectional area is $1.5\,\text{mm}^2$.
2. Calculate the resistance of 50 m of copper cable $4\,\text{mm}^2$ in cross-sectional area.
3. Find the cross-sectional area of a copper cable which is 90 m long and has a resistance of 0.267 Ω.
4. Find the cross-sectional area of a copper cable 42 m long which carries a current of 36 A with a voltage drop of 2.69 V.
5. A certain grade of resistance wire has a resistivity of $50 \times 10^{-8}\,\Omega\,m$. Find the length of this wire needed to make a heating element of resistance 54 Ω. Assume the cross-sectional area of the wire is $0.4\,\text{mm}^2$.
6. Calculate the voltage drop produced in a 75 m length of twin copper cable $16\,\text{mm}^2$ in cross-sectional area when it carries 25 A. What would be the voltage drop if the same size of aluminium cable were used?
7. Determine the resistance of 30 m of copper busbar 60 mm by 6 mm.
8. Calculate the thickness of an aluminium busbar 60 mm wide and 12 m long which has a resistance of 0.000 946 Ω.
9. Find the resistance of 35 m of $1\,\text{mm}^2$ copper cable.

10. Calculate the voltage drop produced by a current of 40 A in 24 m of single 10 mm² copper cable.

11. Find the length of resistance wire 1.2 mm in diameter needed to construct a 20 Ω resistor. (Resistivity $= 50 \times 10^{-8}\,\Omega\,\text{m}$.)

12. Find the resistance of 125 m of 50 mm² aluminium cable.

13. A 6 mm² copper twin cable carries a current of 32 A, and there is a voltage drop of 4.5 V. Calculate the length of the cable.

14. Iron is sometimes used for making heavy-duty resistors. Its resistivity is $12 \times 10^{-8}\,\Omega\,\text{m}$. Calculate the resistance of an iron grid, the effective length of which is 3 m and which is 10 mm by 6 mm in cross-section.

15. Calculate the diameter of an aluminium busbar which is 24 m long and whose resistance is 0.001 39 Ω.

16. A d.c. load current of 28 A is to be supplied from a point 30 m away. Determine a suitable cross-sectional area for the copper cable in order that the voltage drop may be limited to 6 V.

17. Calculate the resistance per 100 m of the following sizes of copper cable:

    (a) 1.5 mm²   (d) 35 mm²
    (b) 6 mm²   (e) 50 mm²
    (c) 10 mm²

18. The resistance of 1000 m of a certain size of cable is given as 0.618 Ω. Find by the method of proportion the resistance of (a) 250 m, (b) 180 m, (c) 550 m.

19. The resistance of a certain length of a cable of cross-sectional area 2.5 mm² is 5.28 Ω. Find by the method of proportion the resistance of a similar length of cable whose cross-sectional area is (a) 10 mm², (b) 25 mm², (c) 1.5 mm².

20. The following figures refer to a certain size of cable:

Length (m)	1000	750	500	250
Resistance (Ω)	0.4	0.3	0.2	0.1

    Plot a graph to show the relationship between resistance and length (length horizontally, resistance vertically) and read from the graph the resistance of a 380 m length of the cable.

21. The following table shows the resistance of cables having the same length but different cross-sectional areas:

c.s.a. (mm²)	1.0	1.5	2.5	4	10
Resistance (Ω)	15.8	12.4	8	3.43	2.29

Plot a graph to show the relationship between resistance and cross-sectional area (cross-sectional area horizontally, resistance vertically). From the graph, find the resistance of a cable whose cross-sectional area is 6 mm^2.

22. The resistance of 100 m of 2.5 mm^2 copper cable is
    (a) 0.712 Ω   (b) 7.12 Ω   (c) 1.404 Ω   (d) 0.0712 Ω

23. The resistance of 15 m of aluminium bar 60 mm by 7.5 mm in cross-section is
    (a) 0.0946 Ω   (c) 9.46 × 10^{-4} Ω
    (b) 9.46 × 10^{-3} Ω   (d) 1.58 × 10^{-4} Ω

24. A shunt for an ammeter is required to have a resistance of 0.002 Ω. If made of copper strip 100 mm long, the cross-sectional area of the strip would be
    (a) 0.89 mm^2 (b) 8.9 mm^2 (c) 1.12 mm^2 (d) 11.2 mm^2

Materials costs, discounts and value-added tax

EXAMPLE 1 The list price of a certain item of equipment is £17.50 but the supplier allows a trade discount of 35%. Calculate the trade price of the item.

$$35\% \text{ of } £17.50 = \frac{35}{100} \times 17.50$$

$$= £6.125$$

∴   trade price $= £17.50 - £6.125$

$$= £11.375$$

EXAMPLE 2 If 50 mm × 50 mm light gauge trunking has a list price of £18.10 per length plus 17.5% VAT, with a trade discount of 35%, calculate the trade price of 14 lengths.

$$\text{List price of 14 lengths} = 14 \times £18.10$$

$$= £253.40$$

Trade discount offered is 35%

$$\therefore \quad \text{discount} = £253.40 \times \frac{35}{100}$$

$$= £88.69$$

$$\text{Trade price of 14 lengths} = £253.40 - £88.69$$

$$= £164.71$$

$$\text{VAT on 14 lengths} = 17.5\% \text{ of } £164.71$$

$$= £164.71 \times \frac{17.5}{100}$$

$$= £28.82$$

$$\text{Cost including VAT} = £164.71 + £28.82$$

$$= £193.53$$

EXAMPLE 3 The retail price of a certain size of masonry drill is £24.52 for 12. Calculate

(a) the cost of each drill,

(b) the cost of 15 drills,

(c) the probable cost of each drill in one year from now with inflation at the rate of 10% per annum.

(a) 12 drills cost £24.52

$$\therefore \quad \text{1 drill costs } \frac{£24.52}{12}$$

$$= £2.04$$

(b) 15 drills would cost

$$15 \times £2.04 = £30.60$$

(c) Probable cost one year from now $= £2.04 + \dfrac{10}{100} \times £2.04$

$$= £2.04 + 20.4\text{p}$$

$$= £2.24$$

EXAMPLE 4   An extension to an existing installation is required, and two types of wiring system are to be considered:
(a) conduit with drawn-in cable,
(b) mineral-insulated cable.
For (a) the materials required are as follows:

12.5 m of conduit	costing	£137 per 90 m
one through box	costing	£1.97
three couplings	costing	£16.35 per 100
three locknuts	costing	£10.23 per 100
two bushes	costing	£36.10 per 100
28 m of cable	costing	£16.24 per 100 m
24 saddles	costing	£42.80 per 100
screws and plugs	costing	£3.20

For (b) the materials required are as follows:

14 m of MIMS cable	costing	£230 per 100 m
2 seals/glands	costing	£21.20 per 10
50 fixing clips	costing	£18.20 per 100
screws and plugs	costing	£3.60

Determine the cost of using each system and state which is the cheaper system from the point of view of the materials used. (Ignore VAT and trade discount.)

(a) Costs are

conduit $\qquad\qquad\qquad\qquad £137 \times \dfrac{12.5}{90} = £19.03$

one through box @ £1.97 $\qquad\qquad\qquad = £1.97$

couplings $\qquad\qquad\qquad\qquad £16.35 \times \dfrac{3}{100} = £0.49$

locknuts $\qquad\qquad\qquad\qquad £10.23 \times \dfrac{3}{100} = £0.31$

male bushes	$£36.10 \times \dfrac{2}{100} = £0.72$
saddles	$£42.80 \times \dfrac{24}{100} = £10.27$
cable	$£16.24 \times \dfrac{28}{100} = £4.55$
screws & plugs @ £3.20	$= \underline{£3.20}$
	Total $\quad £40.54$

(b) Costs are

MIMS cable	$£230 \times \dfrac{14}{100} = £32.20$
seals/glands	$£21.20 \times \dfrac{2}{100} = £0.42$
clips	$£18.20 \times \dfrac{50}{100} = £9.10$
screws/plugs @ £3.60	$= \underline{£3.60}$
	Total $\quad £45.32$

Hence, from the point of view of materials only, the conduit system is the cheaper.

EXERCISE 16

1. If 60 lengths of cable tray cost £732 including VAT, calculate (a) the cost of each length, (b) the cost of 17 lengths.
2. If 66 m of black-enamelled heavy-gauge conduit cost £87 including VAT, calculate (a) the cost per metre, (b) the cost of 245 m.
3. If 400 woodscrews cost £4.52, calculate (a) the cost of 250 screws, (b) the number of screws which could be purchased for £5.28.
4. If 100 m of heavy-gauge plastic conduit is listed at £85.20, plus VAT at 17.5%, calculate the price of 100 m to the customer.
5. The list price of electrician's solder is £360 for 20 kg plus VAT at 17.5% and with a special trade discount of 25%. Calculate the invoice price of 4 kg of solder.
6. An invoice was made out for 20 lengths of 50 mm × 50 mm cable trunking. Each length cost £17.55 plus 17.5% VAT, less 35% trade discount. Calculate the invoice total.

7. An alteration to an existing installation requires the following material:

12 m of plastic trunking at £6.23 per m,

14.5 m of plastic conduit at £86 per 100 m,

45 m of cable at £14.60 per 100 m,

29 single socket-outlets at £12.15 each,

saddles, screws, plugs, etc. £9.20.

Calculate the total cost of the materials. VAT is chargeable at 17.5%.

8. An order was placed one year ago for the following items:

135 m MIMS cable at £217 per 100 m,

500 pot-type seals/glands at £26 per 10,

200 one-hole clips at £29 per 100.

Calculate (a) the original cost of this order; (b) the present cost of the order, allowing 15% per annum for inflation. VAT is chargeable at 17.5% at both (a) and (b).

9. The materials list for an installation is as follows:

45 m of 1.00 mm^2 twin with earth cable at £19.30 per 100 m,

45 m of 2.5 mm^2 twin with earth cable at £28.20 per 100 m,

nine two-gang one-way switches at £3.50 each,

two two-way switches at £2.12 each,

six single switched socket outlets at £3.35 each,

two twin switched socket outlets at £6.40 each,

one consumer unit at £62.20,

sheathing, screws, plugs, etc. £8.00.

Determine the total cost of the materials for this work, adding 17.5% VAT.

10. A contractor's order for conduit and fittings reads as follows:

360 m of 20 mm BEHG steel conduit at £147 per 90 m,

50 20 mm BE standard circular terminal end boxes at £1.81 each,

50 20 mm BE standard through boxes at £2.17 each,

50 20 mm BE standard tee boxes at £2.57 each,

50 20 mm spacer-bar saddles at £23.20 per 100,

50 20 mm steel locknuts at £14.90 per 100,

50 20 mm brass hexagon male bushes at £38 per 100.

All prices are list, the contractor's discount on all items is 40%, and VAT is chargeable at 17.5%. Calculate the invoice total for this order.

11. It is necessary to install six tungsten-halogen flood lighting luminaires outside a factory and the following equipment is required.

Manufacturer's list prices are as shown:

6 off 500 W 'Teck' T/H luminaires	at £14.50 each*,
1 off 'Teck' PIR sensor/relay unit	at £24.10 each*,
80 m 20 mm galvanized steel conduit	at £186 per 100 m,
6 off 20 mm galvanized tee boxes	at £275 per 100,
1 off 20 mm galvanized angle box	at £265 per 100,
7 off galvanized box lids and screws	at £11 per 100,
8 off 20 mm galvanized couplings	at £19 per 100,
30 off 20 mm spacer saddles	at £17.20 per 100,
14 off 20 mm brass male bushes	at £38 per 100,
1 off 'TYLOR' 20 A switch-fuse	at £24.50 each*,
1 off 'TYLOR' 10 A one-way switch	at £3.20 each*,
180 m 1.5 mm^2 pvc single cable	at £12.15 per 100 m,
3 m 0.75 mm^2 three-core pvc flex	at £26.30 per 100 m,
9 off 10 A three-way porcelain connectors	at £80 per 100.

Plus sundries taken from own stock, allow £15.

The wholesaler offers a 25% discount on non-branded items and 10% on branded * items. Calculate (a) the basic cost of the materials and (b) the total cost including VAT at 17.5%.

12. The list prices of certain equipment are as follows:
    (a) £570.30 with 25% discount,
    (b) £886.20 with 40% discount,
    (c) £1357.40 with 10% discount,
    (d) £96.73 with 35% discount.
    For each of the above establish:
    (i) the basic cost price,
    (ii) the VAT chargeable.

13. For each of the following VAT inclusive prices establish the basic cost price:
    (a) £656.25   (d) £1025.27
    (b) £735.33   (e) £3275.72
    (c) £895.43

14. A certain cable is priced at £19.50 per 100 m plus 17.5% VAT. The cost of 65 m is:
    (a) £22.91   (b) £16.09   (c) £14.89   (d) £10.46

15. A certain item of equipment was invoiced at £25.75 and this included VAT at 17.5%. The list price of the item was:
    (a) £3.84   (b) £21.91   (c) £30.26   (d) £43.25

Heating effect

In heating operations

$$\text{heat energy} = \text{mass} \times \begin{bmatrix} \text{temperature} \\ \text{rise} \end{bmatrix} \times \begin{bmatrix} \text{specific heat} \\ \text{capacity} \end{bmatrix}$$

$$Q = m \times (\theta_2 - \theta_1) \times c \text{ joules}$$

where m is the mass of substance in kilograms;

c is its specific heat capacity in joules per kilogram per degree Celsius, J/(kg °C);

θ_1 and θ_2 are respectively the lower and upper temperatures in degrees Celsius.

EXAMPLE 1   Find the heat energy required to raise the temperature of 0.068 m³ of water from 18 °C to 80 °C.
The mass of 1 m³ of water is 10^3 kg, and the specific heat capacity of water is 4187 J/(kg °C).

$$Q = m \times (\theta_2 - \theta_1) \times c$$

$$= 0.068\,\text{m}^3 \left[\frac{10^3\,\text{kg}}{1\,\text{m}^3} \right] \times (80 - 18)\,°\text{C} \times 4187\,\frac{\text{J}}{\text{kg}\,°\text{C}}$$

$$= 0.068 \times 1000 \times 62 \times 4187\,\text{J} \quad \text{(note that the units cancel correctly)}$$

$$= 17\,650\,000\,\text{J} \quad \text{or} \quad 17.65 \times 10^6\,\text{J}$$

Since the joule is also the unit of electrical energy, this is the amount of electrical energy necessary to perform this operation, assuming that no energy is wasted.

EXAMPLE 2   Heat energy of 10 MJ is supplied to 113 litres of water initially at 15 °C. Assuming that none of the heat is wasted, calculate the final temperature of the water.

$$Q = m \times (\theta_2 - \theta_1) \times c$$

$$1\,\mathrm{m}^3 = 10^3\,\mathrm{litres}$$

or $1\,\mathrm{litre} = \dfrac{1}{10^3}\,\mathrm{m}^3$

$\therefore$ $113\,\mathrm{litres} = \dfrac{113}{10^3}\,\mathrm{m}^3$

$$= 0.113\,\mathrm{m}^3$$

$$10 \times 10^6 = 0.113 \times 10^3 \times (\theta_2 - 15) \times 4187$$

$\therefore$ $\theta_2 - 15 = \dfrac{10\,000\,000}{0.113 \times 1000 \times 4187}$

$$= 21.13$$

$\therefore$ $\theta_2 = 21.13 + 15$

$$= 36.13\,°\mathrm{C} \quad \mathrm{or} \quad 36.1\,°\mathrm{C}$$

EXAMPLE 3 Water of volume $0.075\,\mathrm{m}^3$ (75 litres) is to be raised in temperature from $15\,°\mathrm{C}$ to $85\,°\mathrm{C}$ using a $3\,\mathrm{kW}$ heater. Assuming that 20% of the energy is wasted, determine the time that the operation will take.

As in example 1,

$$Q = m \times (\theta_2 - \theta_1) \times c$$

$$= 0.075 \times 10^3 \times (85 - 15) \times 4187$$

$$= 21.98 \times 10^6\,\mathrm{J}$$

A 20% heat loss indicates an efficiency of 80%, and the energy supplied will be

$$Q = 21.98 \times 10^6 \times \frac{100}{80}\,\mathrm{J}$$

$$= 27.5 \times 10^6\,\mathrm{J}$$

The heater supplies energy at the rate of $3\,\mathrm{kW}$, or $3000\,\mathrm{J/s}$. The time required is therefore

$$t = \frac{27.5 \times 10^6}{3000\,\mathrm{J/s}}\,\mathrm{J}$$

$$= 9.166 \times 10^3\,\mathrm{s} \quad \text{(note correct position of units to cancel, leaving seconds)}$$

$$= 2.55\,\mathrm{hours}$$

EXERCISE 17

The mass of 1 m³ of water is 10³ kg, and the specific heat capacity of water is 4187 J/(kg°C). 1 m³ = 10³ litres.

1. Determine the heat energy required to raise the temperature of 100 litres of water from 16 °C to 86 °C. Assume no loss of heat.

2. A tank holds 0.15 m³ of water initially at 24 °C. Determine the heat energy necessary to bring the water to boiling point, assuming no loss of heat.

3. A cold water storage tank contains 120 litres of water. Calculate the heat energy to be supplied to the water to just prevent it from freezing when the surrounding temperature is −2 °C.

4. Energy of 800 000 J is supplied to 0.08 m³ of water initially at 18 °C. Calculate the final temperature, assuming no loss of heat.

5. A heater supplies 25×10^6 J of energy to 0.12 m³ of water initially at 17 °C. If 20% of the heat is wasted, determine the final temperature of the water.

6. A tank 0.5 m square and 0.75 m high is full of water initially at 20 °C. Determine the energy required to raise the temperature of the water to 85 °C, assuming an efficiency of 82% for the operation.

7. Calculate the electrical energy (in kilowatt hours) required to raise the temperature of 0.12 m³ of water through 80 degrees Celsius, neglecting heat losses. ($1 \text{ kWh} = 3.6 \times 10^6$ J.)

8. A tank contains 150 litres of water and is fitted with a 3 kW immersion heater. The initial water temperature is 17 °C. Assuming no loss of heat, determine the water temperature at the end of 2 hours.

9. Determine the time required by a 3 kW immersion heater to raise the temperature of 0.12 m³ of water from 19 °C to 90 °C:
   (a) with no loss of heat,
   (b) if the equipment is 82% efficient.

10. Determine the rating of a heating element which will raise the temperature of 0.2 m³ of water through 80 degrees Celsius in $4\frac{1}{2}$ hours. The efficiency of the operation is 80%.

11. Determine the cost of raising 50 litres of water to boiling point in a 3.5 kW washing machine which is 78% efficient. The initial water temperature is 18 °C. Electrical energy costs 5.36p per kWh. How long will the operation take?

12. A test on an electric kettle yielded the following results:

Volume of water	0.001 m³	Initial temperature	18 °C
e.m.f.	235 V	Final temperature	100 °C
Current	6.2 A	Time taken	330 s

Determine the efficiency of the kettle.

13. A water heater containing 12 litres of water is fitted with a 750 W element rated at 250 V and may be assumed to be 82% efficient. Determine the time required for it to raise the temperature of the water through 45 degrees Celsius:

    (a) at normal voltage;
    (b) if the voltage falls to 230 V.

14. A tank contains 0.15 m³ of water and is fitted with a 240 V, 3 kW heater. Heat losses from the tank account for 17% of the energy supplied. Due to voltage drop, the actual voltage at the heater is 235 V. Determine the time required to raise the temperature of the water through 52 degrees Celsius.

15. A tank is 0.5 m by 0.75 m × 0.75 m and is filled with water initially at 18 °C. Calculate the rating (in kW) of a heater which will raise the temperature of the water to 82 °C in 2 hours. The efficiency may be taken as 78%. Determine also (a) the cost of the operation if electrical energy costs 5.18p per kWh, (b) the resistance of the element to work from the 240 V supply.

16. A cylindrical tank, 0.5 m in diameter and 1 m high, is fitted with a 3 kW heater. The tank is full of water initially at 18 °C, and heat losses amount to 18% of that supplied. Determine the time taken to raise the temperature of the water to 85 °C.

17. A 1000 W kettle which is 84% efficient is filled with 1 litre of water at 16 °C. What will the water temperature be after 4 minutes? Assume that the kettle is supplied at its rated voltage.

18. A certain industrial process requires 0.25 m³ of water per hour at 86 °C. Assuming that the initial temperature of the water is 15 °C and that the efficiency of the boiler is 76%, calculate

    (a) the current taken by the heater from a 240 V supply,
    (b) the resistance of the element,
    (c) the cost per hour with energy at 5.48p per kWh.

19. Determine the efficiency of an electric washing machine from the following test results:

Water 36 litres	e.m.f. 240 V
Initial temperature 18 °C	Current 14.5 A
Final temperature 100 °C	Time taken 1.25 hours

20. Make a sketch showing the working parts of a thermostat-controlled electric water heater of the free-outlet type.

    A water heater of this type rated at 1000 W, 240 V contains 0.015 m^3 of water. Calculate the time required to raise the temperature of the water from 20 °C to 85 °C if the overall efficiency is 84%.

21. The amount of the heat energy required to raise the temperature of 0.08 m^3 (80 litres) of water from 19 °C to 85 °C is

    (a) 1.26 J
    (c) 22 107 360 J
    (b) 22.107 × 10^9 J
    (d) 22 107 J

22. A 1000 W kettle contains 1 litre of water at 15 °C. It is 85% efficient. The time taken for the water to boil is

    (a) 5.1 min (b) 0.6 min (c) 6.0 min (d) 7.0 min

Electromagnetic effect

MAGNETIC FLUX AND FLUX DENSITY

The unit of magnetic flux is the *weber* (Wb). A magnetic field has a value of 1 Wb if a conductor moving through it in one second has an e.m.f. of 1 V induced in it.

Convenient units often used are the milliweber (mWB)

$$1 \, \text{Wb} = 10^3 \, \text{mWb}$$

and the microweber (μWb)

$$1 \, \text{Wb} = 10^6 \, \text{μWB}$$

The symbol for magnetic flux is Φ.

The total flux divided by the cross-sectional area of the magnetic field (taken at right angles to the direction of the flux) gives the flux density (symbol B), thus

$$B = \frac{\Phi}{A}$$

where Φ is the total flux (Wb),

        A is the cross-sectional area of the field (m^2),

and B is the flux density (Wb/m^2 or tesla, T).

EXAMPLE 1   The total flux in the air gap of an instrument is 0.15 mWb and the cross-sectional area of the gap is 500 mm^2. Calculate the flux density.

$$B = \frac{\Phi}{A}$$

Convert 0.15 mWb to Wb:

$$0.15\,mWb = 0.15\,\cancel{mWb} \left[\frac{1\,Wb}{1000\,\cancel{mWb}} \right]$$

$$= \frac{0.15}{1000}\,Wb$$

Then $B = \dfrac{0.15\,Wb}{1000 \times 500/10^6\,m^2}$ (note conversion of mm^2 to m^2)

$$= \frac{0.15 \times 1\,000\,000}{1000 \times 500}$$

$$= 0.3\,Wb/m^2 \quad \text{or} \quad 0.3\,T$$

EXAMPLE 2   The air gap of a contactor is 15 mm square. The flux density is 1.2 T. Calculate the total flux.

$$B = \frac{\Phi}{A}$$

$\therefore$ $\Phi = B \times A$

$$A = 15\,\cancel{mm} \times \left[\frac{1\,m}{1000\,\cancel{mm}} \right] \times 15\,\cancel{mm} \times \left[\frac{1\,m}{1000\,\cancel{mm}} \right]$$

$$= \frac{225}{10^6}\,m^2$$

$$\therefore \quad \Phi = 1.2\frac{\text{Wb}}{\text{m}^2} \times \frac{225}{10^6}\,\text{m}^2$$

$$= \frac{270}{10^6}\,\text{Wb}$$

Changing to more convenient units,

$$\Phi = \frac{270}{10^6}\,\text{Wb}\left[\frac{1000\,\text{mWb}}{1\,\text{Wb}}\right]$$

$$= 0.27\,\text{mWb}$$

FORCE ON A CURRENT-CARRYING CONDUCTOR IN A MAGNETIC FIELD

The force on a current-carrying conductor situated in and at right angles to a magnetic field is given by

$$F = BlI$$

where F is the force in newtons (N),

B is the flux density (T),

l is the effective conductor length (m),

and I is the current (A).

EXAMPLE   A conductor 0.25 m long situated in and at right angles to a magnetic field experiences a force of 5 N when the current through it is 50 A. Determine the flux density.

$$F = B \times l \times I$$

$$5 = B \times 0.25 \times 50$$

$$\therefore \quad B = \frac{5}{0.25 \times 50}$$

$$= 0.4\,\text{T}$$

INDUCED E.M.F.

When a conductor moves through a magnetic field in a direction at right angles to the field, the e.m.f. induced is

$$e = Blv \text{ volts}$$

where B is the flux density (T),

        l is the effective length of conductor (m),

and v is the velocity of the conductor (m/s).

EXAMPLE I   The e.m.f. induced in a conductor of effective length 0.25 m moving at right angles through a magnetic field at a velocity of 5 m/s is 1.375 V. Calculate the magnetic flux density.

$$e = Blv$$

$$\therefore \quad 1.375 = B \times 0.25 \times 5$$

$$\therefore \quad B = \frac{1.375}{0.25 \times 5}$$

$$= 1.1\,\text{T}$$

If the flux linking a coil of N turns is Φ_1 (Wb) at some instant of time t_1 and Φ_2 (Wb) at another instant of time t_2, the e.m.f. induced is

$$e = \text{number of turns} \times \text{rate of change of flux}$$

$$= N \times \frac{\text{change in flux}}{\text{time required for change in flux}}$$

$$= N \times \frac{\Phi_2 - \Phi_1}{t_2 - t_1}\,\text{volts}$$

EXAMPLE 2   The flux linking a coil of 50 turns changes from 0.042 Wb to 0.075 Wb in 0.003 seconds. Calculate the e.m.f. induced.

$$e = N \times \frac{\Phi_2 - \Phi_1}{t_2 - t_1}$$

Here $t_2 - t_1 = 0.003$

thus $e = 50 \times \dfrac{0.075 - 0.042}{0.003}$

$$= 550\,\text{V}$$

SELF-INDUCTANCE

If the self-inductance of a magnetic system is L henrys and the current changes from I_1 at time t_1 to I_2 at time t_2, the induced e.m.f. is

$$e = L \times \text{rate of change of current}$$

$$= L \times \frac{I_2 - I_1}{t_2 - t_1} \text{ volts}$$

where the current is in amperes and the time in seconds.

EXAMPLE 1   A coil has self-inductance 3 H, and the current through it changes from 0.5 A to 0.1 A in 0.01 s. Calculate the e.m.f. induced.

$$e = L \times \text{rate of change of current}$$

$$= 3 \times \frac{0.5 - 0.1}{0.01}$$

$$= 120 \, \text{V}$$

The self-inductance of a magnetic circuit is given by

$$\text{self-inductance} = \frac{\text{change in flux linkage}}{\text{corresponding change in current}}$$

$$L = N \times \frac{\Phi_2 - \Phi_1}{I_2 - I_1} \text{ henrys}$$

where N is the number of turns on the magnetizing coil and Φ_2, I_2; Φ_1, I_1 are corresponding values of flux and current.

EXAMPLE 2   The four field coils of a d.c. machine each have 1250 turns and are connected in series. The change in flux produced by a change in current of 0.25 A is 0.0035 Wb. Calculate the self-inductance of the system.

$$L = N \times \frac{\Phi_2 - \Phi_1}{I_2 - I_1}$$

$$= 4 \times 1250 \times \frac{0.0035}{0.25}$$

$$= 70 \, \text{H}$$

MUTUAL INDUCTANCE

If two coils A and B have mutual inductance M henrys, the e.m.f. in coil A due to current change in coil B is

$$e_A = M \times \text{rate of change of current in coil B}$$

Thus, if the current in coil B has values I_1 and I_2 at instants of time t_1 and t_2,

$$e = M \times \frac{I_2 - I_1}{t_2 - t_1} \text{volts}$$

EXAMPLE 1 Two coils have mutual inductance 3 H. If the current through one coil changes from 0.1 A to 0.4 A in 0.15 s, calculate the e.m.f. induced in the other coil.

$$e = 3 \times \frac{0.4 \times 0.1}{0.15} (t_2 - t_1 = 0.15)$$

$$= 6\,V$$

The mutual inductance between two coils is given by

$$M = N_A \times \frac{\Phi_2 - \Phi_1}{I_{B1} - I_{B2}} \text{henrys}$$

where N_A is the number of turns on coil A and Φ_2 and Φ_1 are the values of flux linking coil A due to the two values of current in coil B, I_{B2} and I_{B1} respectively.

EXAMPLE 2 The secondary winding of a transformer has 200 turns. When the primary current is 1 A the total flux is 0.05 Wb, and when it is 2 A the total flux is 0.095 Wb. Assuming that all the flux links both windings, calculate the mutual inductance between the primary and secondary.

$$M = N_A \times \frac{\Phi_2 - \Phi_1}{I_{B1} - I_{B2}}$$

$$= 200 \times \frac{0.095 - 0.05}{2 - 1}$$

$$= 9\,H$$

1. Convert (a) 0.001 25 Wb to milliwebers, (b) 795 000 μWb to webers.

2. Complete the following table:

Wb	0.025		0.74	
mWb		35		
μWb			59 500	850 000

3. The flux density in an air gap of cross-sectional area 0.0625 m² is 1.1 T. Calculate the total flux.

4. Determine the flux density in an air gap 120 mm by 80 mm when the total flux is 7.68 mWb.

5. An air gap is of circular cross-section 40 mm in diameter. Find the total flux when the flux density is 0.75 T.

6. Calculate the force on a conductor 150 mm long situated at right angles to a magnetic field of flux density 0.85 T and carrying a current of 15 A.

7. Determine the flux density in a magnetic field in which a conductor 0.3 m long situated at right angles and carrying a current of 15 A experiences a force of 3.5 N.

8. Complete the table below, which relates to the force on conductors in magnetic fields:

Flux density (T)	0.95	0.296	1.2	0.56	
Conductor length (m)	0.035		0.3	0.071	0.5
Current (A)		4.5		0.5	85
Force (N)	0.05	0.16	12		30

9. A conductor 250 mm long is situated at right angles to a magnetic field of flux density 0.8 T. Choose six values of current from 0 to 5 A, calculate the force produced in each case, and plot a graph showing the relationship between force and current.

10. If the conductor of question 9 is to experience a constant force of 1.5 N with six values of flux density ranging from 0.5 T to 1.0 T, calculate the current required in each case and plot a graph showing the relationship between current and flux density.

11. A conductor 250 mm long traverses a magnetic field of flux density 0.8 T at right angles. Choose six values of velocity from 5 to 10 m/s. Calculate the induced e.m.f. in each case and plot a graph of e.m.f. against velocity.

12. If the conductor of question 11 is to experience a constant induced e.m.f. of 3 V with values of flux density varying from 0.5 T to 1.0 T,

choose six values of flux density, calculate in each case the velocity required, and plot a graph of velocity against flux density.

13. A conductor of effective length 0.2 m moves through a uniform magnetic field of density 0.8 T with a velocity of 0.5 m/s. Calculate the e.m.f. induced in the conductor.

14. Calculate the velocity with which a conductor 0.3 m long must pass at right angles through a magnetic field of flux density 0.65 T in order that the induced e.m.f. shall be 0.5 V.

15. Calculate the e.m.f. induced in a coil of 1200 turns when the flux linking with it changes from 0.03 Wb to 0.045 Wb in 0.1 s.

16. The magnetic flux in a coil of 850 turns is 0.015 Wb. Calculate the e.m.f. induced when this flux is reversed in 0.25 s.

17. A coil has self-inductance 0.65 H. Calculate the e.m.f. induced in the coil when the current through it changes at the rate of 10 A/s.

18. A current of 5 A through a certain coil is reversed in 0.1 s, and the induced e.m.f. is 15 V. Calculate the self-inductance of the coil.

19. A coil has 2000 turns. When the current through the coil is 0.5 A the flux is 0.03 Wb; when the current is 0.8 A the flux is 0.045 Wb. Calculate the self-inductance of the coil.

20. A air-cored coil has 250 turns. The flux produced by a current of 5 A is 0.035 Wb. Calculate the self-inductance of the coil. (Hint: in an air-cored coil, current and magnetic flux are directly proportional. When there is no current, there is no flux.)

21. Two coils have mutual inductance 2 H. Calculate the e.m.f. induced in one coil when the current through the other changes at the rate of 25 A/s.

22. Two coils have mutual inductance 0.15 H. At what rate must the current through one change in order to induce an e.m.f. of 10 V in the other?

23. Two coils are arranged so that the same flux links both. One coil has 1200 turns. When the current through the other coil is 1.5 A, the flux is 0.045 Wb; when the current is 2.5 A the flux is 0.07 Wb. Calculate the mutual inductance between the coils.

24. Calculate the e.m.f. induced in one of the coils of question 23 if a current of 0.2 A in the other coil is reversed in 0.15 s.

25. The total magnetic flux in an air gap is given as 200 μW. In milliwebers this is

(a) 0.2   (b) 20   (c) 0.02   (d) 2

26. the total flux in a magnetic circuit is 0.375 mWb and the cross-sectional area is 5 cm^2. The flux density is

(a) 1.333 T   (b) 0.075 T   (c) 0.75 T   (d) 7.5 T

27. A force of 0.16 N is experienced by a conductor 500 mm long carrying a current of 0.375 A and resting at right angles to a uniform magnetic field. The magnetic flux density is

(a) 0.117 T   (b) 0.85 T   (c) 8.5 T   (d) 0.085 T

28. The e.m.f. induced in a conductor of length 0.15 m moving at right angles to a magnetic field with a velocity of 7.5 m/s is 22.5 mV. The magnetic flux density is

(a) 20 T   (b) 25.3 T   (c) 0.02 T   (d) 0.0253 T

29. The magnetic flux linking a coil of 150 turns changes from 0.05 Wb to 0.075 Wb in 5 ms. The e.m.f. induced is

(a) 750 V   (b) 0.75 V   (c) 37.5 V   (d) 37 500 V

30. When the current through a coil changes from 0.15 A to 0.7 A in 0.015 s, the e.m.f. induced is 100 V. The self-inductance of the coil is

(a) 367 H   (b) 0.367 H   (c) 2.73 H   (d) 1.76 H

31. Two coils have mutual inductance 0.12 H. The current through one coil changes at the rate of 150 A/s. The e.m.f. induced in the other is

(a) 1250 V   (b) 0.0008 V   (c) 180 V   (d) 18 V

Alternating e.m.f. and current, power and power factor

ALTERNATING E.M.F. AND CURRENT

The value and direction of the e.m.f. induced in a conductor rotating at constant speed in a uniform magnetic field, Fig. 25(a), vary according to the position of the conductor.

The e.m.f. can be represented by the displacement QP of the point P above the axis XOX, Fig. 25(b). OP is a line which is rotating about the point O at the same speed as the conductor is rotating in the magnetic field. The length of OP represents the maximum value of the induced voltage. OP is called a *phasor*.

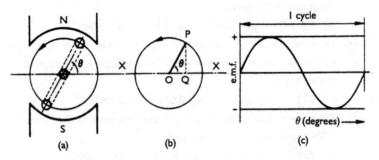

Fig. 25

A graph, Fig. 25(c), of the displacement of the point P plotted against the angle θ (the angle through which the conductor has moved from the position of zero induced e.m.f.) is called a *sine wave*, since PQ is proportional to the sine of angle θ. One complete revolution of OP is called a *cycle*.

EXAMPLE I   An alternating voltage has a maximum value of 200 V. Assuming that it is sinusoidal in nature (i.e. it varies according to a sine wave), plot a graph to show the variations in this voltage over a complete cycle.
Method (Fig. 26) Choose a reasonable scale for OP; for instance, 10 mm ≡ 100 V.

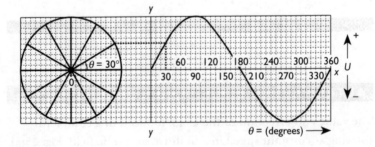

Fig. 26

Draw a circle of radius 20 mm at the left-hand side of a piece of graph paper to represent the rotation of OP.

One complete revolution of OP sweeps out 360°. Divide the circle into any number of equal portions, say 12. Each portion will then cover 30°.

Construct the axes of the graph, drawing the horizontal axis OX (the x-axis) on a line through the centre of the circle. This x-axis should now be marked off in steps of 30° up to 360°. If desired, perpendicular lines can be drawn through these points. Such lines are called *ordinates*.

The points on the graph are obtained by projecting from the various positions of P to the ordinate corresponding to the angle θ at that position.

Remember that when $\theta = 0°$ and 180° the generated e.m.f. is zero, and when $\theta = 90°$ and 270° the generated e.m.f. has its maximum value.

EXAMPLE 2   Determine the average and r.m.s. values of the voltage generated in example 1.

For this purpose a larger graph is desirable, but only one half of the cycle need be shown (Fig. 27).

The base of the graph must now be divided into any number of equal portions. We can use the 30° plotting points to do this. At the centre of each portion of the graph so formed, a *mid-ordinate* is then erected.

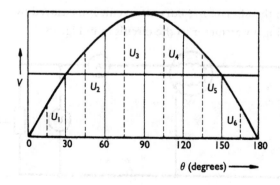

Fig. 27

Calling these mid-ordinates u_1, u_2, u_3, etc., the *average* value of the voltage is

$$U_{\text{av.}} = \frac{u_1 + u_2 + u_3 + \ldots + u_n}{n}$$

where n is the number of mid-ordinates.

In this case,

$$U_{\text{av.}} = \frac{52 + 140 + 190 + 190 + 140 + 52}{6}$$

$$= 127\,\text{V}$$

The *root-mean-square* or r.m.s. value is

$$U_{\text{r.m.s.}} = \sqrt{\frac{u_1^2 + u_2^2 + u_3^2 + \ldots + u_n^2}{n}}$$

$$= \sqrt{\frac{52^2 + 140^2 + 190^2 + 190^2 + 140^2 + 52^2}{6}}$$

$$= 140\,\text{V}$$

Actually, for a pure sine wave,

$$U_{\text{av.}} = 0.637 \times U_{\text{max.}}$$

and $\quad U_{\text{r.m.s.}} = 0.707 \times U_{\text{max.}}$

and the same relationships hold good for maximum, average, and r.m.s. values of current.

POWER IN THE ALTERNATING-CURRENT CIRCUIT

Power in the alternating-current circuit is measured by connecting a wattmeter in the circuit as in Fig. 28.

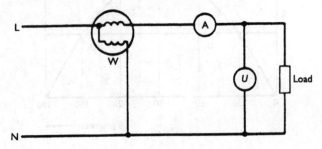

Fig. 28

If an ammeter and voltmeter are also included as shown, this enables the *power factor* to be measured:

if U is the supply voltage (voltmeter reading),

       I is the current (ammeter reading),

and P is the power (wattmeter reading),

then $P = U \times I \times$ power factor (p.f.)

EXAMPLE 1   An a.c. motor takes 7.5 A from a 230 V supply and a wattmeter connected in the circuit shows 1380 watts. Calculate the power factor.

$$P = U \times I \times \text{p.f.}$$

$\therefore$ $1380 = 230 \times 7.5 \times \text{p.f.}$

$\therefore$ $\text{p.f.} = \dfrac{1380}{230 \times 7.5}$

            $= 0.8$

(There are no units: power factor is a number which is never greater than 1.)

EXAMPLE 2   The current supplied to a fluorescent circuit is 0.68 A at a voltage of 230 V and power factor 0.77. Calculate the power supplied.

$P = U \times I \times \text{p.f.}$

   $= 230 \times 0.68 \times 0.77$

   $= 120.4\,\text{W}$

EXERCISE 19

1. Plot a sine wave, over one complete cycle, of an alternating voltage having a maximum value of 340 V. Determine the r.m.s. value of this voltage.

2. An alternating current has the following values taken at intervals of 30° over one half cycle:

Angle (θ)	0°	30°	60°	90°	120°	150°	180°
Current (A)	0	10.5	17.5	19.7	15.0	11.5	0

Determine the average and r.m.s. values of this current.

3. Plot a graph over one complete cycle of a sinusoidal alternating voltage having a r.m.s. value of 200 V.

4. Find the average and r.m.s. values of a current whose waveform is triangular, having zero values at $0°$ and $180°$ and a maximum value of 50 mA at $90°$.

5. An e.m.f. has the following values taken at $30°$ intervals over one half cycle:

Angle (θ)	0°	30°	60°	90°	120°	150°	180°
e.m.f. (V)	0	16	44	100	44	16	0

The e.m.f. is applied to a 100 Ω resistor for 2 minutes. Determine

(a) the average and r.m.s. values of the resulting current,

(b) the power,

(c) the energy supplied.

6. A square-wave current has a value of 25 mA from $0°$ to $30°$, a zero value from $30°$ to $60°$, and a value of 25 mA from $60°$ to $90°$. It is led into the same conductor as the current of question 4. By adding corresponding instantaneous values together, plot a graph of the resulting current for one half cycle.

7. Complete the following table:

Voltage (V)	240	238	242	
Current (A)	3	5		4.8
Power (W)		893	1382	338
Power factor	0.8		0.56	0.64

8. An a.c. motor takes a current of 9.8 A from a 240 V supply at a power factor of 0.8. The power absorbed is

(a) 0.0326 W (b) 19.6 W (c) 1882 W (d) 2940 W

9. A fluorescent luminaire takes a current of 1.75 A at 238 V and the power absorbed is 210 W. The power factor is

(a) 1.98 (b) 0.5 (c) 0.65 (d) 1.54

10. A circuit takes 546 W from a supply at 235 V at a power factor of 0.65. The current is

(a) 3.58 A (b) 0.28 A (c) 197 400 A (d) 0.662 A

Transformer ratios

$$\frac{U_p}{U_s} = \frac{N_p}{N_s} = \frac{I_s}{I_p}$$

U_p is the primary voltage,

I_p is the primary current,

N_p is the number of turns on the primary winding,

U_s is the secondary voltage,

I_s is the secondary current,

and N_s is the number of turns on the secondary winding.

A step-down transformer is one in which the secondary *voltage* is less than the primary *voltage*.

EXAMPLE I   A transformer has a step-down ratio of 20:1 (i.e. $U_p/U_s = 20/1$). Its primary winding consists of 2000 turns. Calculate

(a) the number of secondary turns,

(b) the secondary voltage when the primary is supplied at 230 V.

(a) $$\frac{U_p}{U_s} = \frac{N_p}{N_s}$$

$$\frac{20}{1} = \frac{2000}{N_s}$$

∴ $$\frac{1}{20} = \frac{N_s}{2000}$$

∴ $$N_s = \frac{2000}{20}$$

$$= 100 \text{ turns}$$

(b) Since this is a step-down transformer, the voltage is reduced in the ratio 20:1,

∴ $$\text{secondary voltage} = \frac{1}{20} \times 230$$

$$= 11.5 \text{ V}$$

EXAMPLE 2 A single-phase transformer, 230/50 V, supplies 150 A from its secondary side. Calculate the primary current.

$$\frac{U_p}{U_s} = \frac{I_s}{I_p}$$

$$\frac{230}{50} = \frac{150}{I_p}$$

$$\therefore \quad \frac{50}{230} = \frac{I_p}{150}$$

$$\therefore \quad I_p = \frac{50 \times 150}{230}$$

$$= 32.6\,\text{A}$$

EXERCISE 20

1. A 230/50 V step-down transformer is supplying a current of 80 A. Calculate the primary current.

2. A transformer is required to supply 100 V from the 230 V mains. Its primary winding contains 2000 turns. Calculate
   (a) the number of secondary turns,
   (b) the primary current when the transformer delivers a current of 15 A.

3. The transformation ratio of a transformer is 6 : 1. Calculate the secondary voltage when the primary voltage is 400 V.

4. Describe the operation of a single-phase transformer, explaining clearly the effects of the various parts.
   The primary winding of a step-down single-phase transformer takes a current of 22 A at 3300 V when working at full load. If the step-down ratio is 14 : 1, calculate the secondary voltage and current.

5. Make a neat diagram or sketch of a simple single-phase double-wound transformer and with its aid explain the action of the transformer.
   Calculate the respective number of turns in each winding of such a transformer which has a step-down ratio of 3040 to 230 V if the 'volts per turn' are 1.6.

6. A transformer has 1920 primary turns and 96 secondary turns. If the primary voltage is 240 V, the secondary voltage is
   (a) 120 V (b) 480 V (c) 12 V (d) 4800 V

7. The secondary current of the transformer of question 6 is 1.25 A. The primary current is

    (a) 2.5 A   **(b)** 25 A   **(c)** 0.625 A   **(d)** 0.0625 A

8. A step-down transformer has a turns ratio of 5 to 1. The number of secondary turns is 150. The number of primary turns is

    (a) 30   **(b)** 7500   **(c)** 750   **(d)** 75

Mechanics

MOMENT OF FORCE

A force F newtons applied at right angles to a rod of length l metres pivoted at P (Fig. 29) produces a turning moment M, where

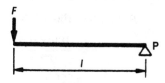

Fig. 29

$$M = F \times l \text{ newton metres (N m)}$$

(Note that this turning moment is produced whether or not the bar is actually free to turn.)

EXAMPLE I   A horizontal bar 0.5 m long is arranged as in Fig. 29. Calculate the force required in order to produce a moment of 250 N m.

$$M = F \times l$$

$$\therefore \quad 250\,\text{N m} = F \times 0.5\,\text{m}$$

$$\therefore \quad F = \frac{250\,\text{N m}}{0.5\,\text{m}}$$

$$= 500\,\text{N}$$

EXAMPLE 2   A horizontal bar 0.75 m long is pivoted at a point 0.5 m from one end, and a downward force of 100 N is applied at right angles to this end of the bar. Calculate the downward force which must be applied at right angles to the other end in order to maintain the bar in a horizontal position. Neglect the weight of the bar.

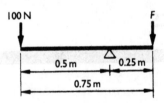

Fig. 30

The principle of moments applies; that is, for equilibrium (see Fig. 30),

$$\text{total clockwise moment} = \text{total anticlockwise moment}$$

$$F \times 0.25 = 100 \times 0.5$$

$$\therefore \quad F = \frac{100 \times 0.5}{0.25}$$

$$= 200\,\text{N}$$

FORCE RATIO

If the bar of example 2 is considered as a lever, then an *effort* of 100 N is capable of exerting a force of 200 N on an object. The force *F* is then in fact the *load*.

By the principle of moments,

$$\text{load} \times \begin{bmatrix} \text{its distance from} \\ \text{the pivot or fulcrum} \end{bmatrix} = \text{effort} \times \begin{bmatrix} \text{its distance} \\ \text{from the pivot} \end{bmatrix}$$

The *force ratio* is $\dfrac{\text{load}}{\text{effort}}$,

i.e. $\quad$ force ratio $= \dfrac{\text{load}}{\text{effort}} = \dfrac{\text{distance from \textit{effort} to pivot}}{\text{distance from \textit{load} to pivot}}$

In the case of example 2,

$$\text{force ratio} = \frac{0.5 \, \cancel{m}}{0.25 \, \cancel{m}}$$

$$= 2$$

Note that force ratio is often also referred to as 'mechanical advantage'.

MASS, FORCE AND WEIGHT

Very often the load is an object which has to be raised to a higher level against the force of gravity.

The force due to gravity acting on a mass of 1 kg is 9.81 N. The force to raise a mass of 1 kg against the influence of gravity is therefore 9.81 N, and this is called the weight of the 1 kg mass.

Although the newton is the correct scientific unit of force and weight, for industrial and commercial purposes it is usual to regard a mass of 1 kg as having a weight of 1 kilogram force (kgf), therefore

$$1 \, \text{kgf} = 9.81 \, \text{N}$$

The kilogram force is the 'gravitational' unit of weight and is often abbreviated to 'kilogram', or even 'kilo', in common usage.

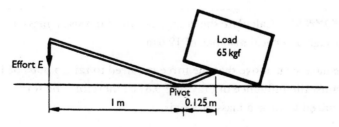

Fig. 31

EXAMPLE A crowbar is arranged as shown in Fig. 31 and for practical purposes the formula for force ratio may be applied to find the effort required to raise a load of 65 kgf:

$$\frac{\text{load}}{\text{effort}} = \frac{\text{distance from effort to pivot}}{\text{distance from load to pivot}}$$

$$\therefore \qquad \frac{65}{E} = \frac{1\,\text{m}}{0.125\,\text{m}}$$

$$\therefore \qquad \frac{E}{65} = \frac{0.125}{1}$$

$$\therefore \qquad E = 65 \times 0.125$$

$$= 8.125\,\text{kgf}$$

WORK

When a force F newtons produces displacement of a body by an amount s metres in the direction of the force, the work done is

$$W = F \times s \text{ newton metres or joules (J)}$$

This is also the energy expended in displacing the body.

EXAMPLE I A force of 200 N is required to move an object through a distance of 3.5 m. Calculate the energy expended.

$$W = F \times s$$
$$= 200\,\text{N} \times 3.5\,\text{m}$$
$$= 700\,\text{N m} \quad \text{or} \quad 700\,\text{J}$$

EXAMPLE 2 Calculate the energy required to raise a mass of 5 kg through a vertical distance of 12.5 m.

We have seen above that the force required to raise a mass of 1 kg against the influence of gravity is 9.81 N; therefore the force required to raise a mass of 5 kg is

$$F = 5 \times 9.81\,\text{N}$$

and the energy required is

$$W = 5 \times 9.81\,\text{N} \times 12.5\,\text{m}$$
$$= 613\,\text{N\,m} \quad \text{or} \quad 613\,\text{J}$$

THE INCLINED PLANE

Figure 32 illustrates a method of raising a load G through a vertical distance h by forcing it up a sloping plane of length l using an effort E.

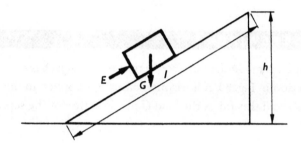

Fig. 32

Ignoring the effects of friction (which can be reduced by using rollers under the load),

$$\begin{bmatrix} \text{energy expended} \\ \text{by the effort} \end{bmatrix} = \begin{bmatrix} \text{energy absorbed} \\ \text{by the load} \end{bmatrix}$$

that is,

$$\text{effort} \times \begin{pmatrix} \text{distance through which} \\ \text{the effort is exerted} \end{pmatrix} = \text{load} \times \begin{pmatrix} \text{vertical distance} \\ \text{through which} \\ \text{the load is raised} \end{pmatrix}$$

$$E \times l = G \times h$$

$$\text{Force ratio} = \frac{\text{load}}{\text{effort}} = \frac{G}{E} = \frac{l}{h}$$

EXAMPLE A motor weighing 100 kgf is to be raised through a vertical distance of 2 m by pushing it up a sloping ramp 5 m long. Ignoring the effects of friction, determine the effort required.

$$\frac{G}{E} = \frac{l}{h}$$

$$\frac{100}{E} = \frac{5\,\cancel{m}}{2\,\cancel{m}}$$

$$\frac{E}{100} = \frac{2}{5}$$

$$E = 100 \times \frac{2}{5}$$

$$= 40\,\text{kgf}$$

THE SCREWJACK

A simplified arrangement of a screw type of lifting jack is shown in cross-section in Fig. 33. A horizontal effort E is applied to the arm of radius r and this raises the load G by the action of the screw thread T.

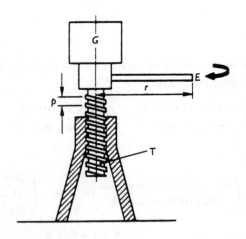

Fig. 33

If the effort is taken through a complete revolution, it acts through a distance equal to $2\pi \times r$ and the load rises through a vertical distance equal to the 'pitch' of the screw thread, which is the distance between successive turns of the thread.

If p is the pitch of the thread, and ignoring friction,

$$\begin{bmatrix} \text{energy expended by} \\ \text{the effort} \end{bmatrix} = \begin{bmatrix} \text{energy absorbed by the load} \\ \text{in rising through a distance } p \end{bmatrix}$$

$$E \times 2\pi r = G \times p$$

The force ratio is

$$\frac{\text{load}}{\text{effort}} = \frac{G}{E} = \frac{2\pi r}{p}$$

EXAMPLE If the pitch of the thread of a screwjack is 1 cm and the length of the radius arm is 0.5 m, find the load which can be raised by applying a force of 20 kg.

$$\frac{G}{E} = \frac{2\pi r}{p}$$

$$\therefore \quad \frac{G}{20} = \frac{2\pi \times 0.5\,\text{m}}{1/100\,\text{m}} \quad \begin{array}{l}\text{(note the conversion} \\ \text{of cm to m)}\end{array}$$

$$\therefore \quad G = \frac{20 \times 2\pi \times 0.5}{0.01}$$

$$= 6283\,\text{kgf}$$

THE WHEEL-AND-AXLE PRINCIPLE

Figure 34 shows a simplified version of a common arrangement by means of which a load G is raised by applying an effort E.

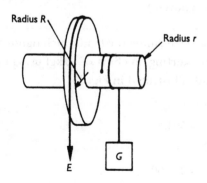

Radius R

Radius r

E

G

Fig. 34

By the principle of moments,

$$E \times R = G \times r$$

$$\therefore \quad \text{force ratio} = \frac{\text{load}}{\text{effort}} = \frac{G}{E} = \frac{R}{r}$$

EXAMPLE Calculate the effort required to raise a load of 250 kgf using the arrangement shown in Fig. 34, if the radius of the large wheel is 20 cm and the radius of the axle is 8 cm.

$$\frac{G}{E} = \frac{R}{r}$$

$$\therefore \quad \frac{250}{E} = \frac{20 \, \text{cm}}{8 \, \text{cm}}$$

$$\therefore \quad \frac{E}{250} = \frac{8}{20}$$

$$\therefore \quad E = 250 \times \frac{8}{20}$$

$$= 100 \, \text{kgf}$$

THE BLOCK AND TACKLE

When a system of forces is in *equilibrium*, the sum of all forces acting downwards is equal to the sum of all forces acting upwards.

Figures 35(a), (b), (c), and (d) illustrate various arrangements of lifting tackle (rope falls) raising a load G by exerting an effort E. In each case the effort is transmitted throughout the lifting rope, giving rise to increasing values of force ratio. (The effects of friction are ignored.)

EXAMPLE Determine the load which (ignoring friction) could be raised by exerting an effort of 50 kgf using each of the arrangements illustrated in Fig. 35.

For (a), $G = E$

$$= 50 \, \text{kgf}$$

For (b) $G = 2E$

$$= 2 \times 50$$

$$= 100 \, \text{kgf}$$

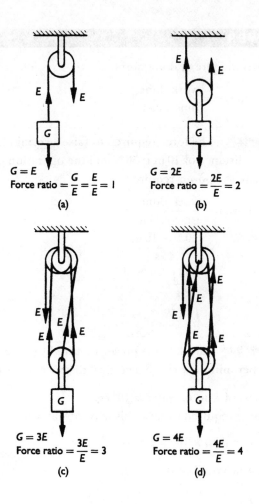

Fig. 35

For (c), $G = 3E$

$\qquad = 3 \times 50$

$\qquad = 150\,\text{kgf}$

For (d), $G = 4E$

$\qquad = 4 \times 50$

$\qquad = 200\,\text{kgf}$

POWER

Power is the rate of doing work,

i.e. $\quad \text{power} = \dfrac{\text{work done}}{\text{time taken}}$

EXAMPLE 1 The force required to raise a certain load through a vertical distance of 10 m is 50 N and the operation takes 45 s. Calculate the power required.

$$\text{Power} = \frac{\text{work done}}{\text{time taken}}$$

$$= \frac{50\,\text{N} \times 10\,\text{m}}{45\,\text{s}}$$

$$= \frac{500\,\text{J}}{45\,\text{s}}$$

$$= 11.1\,\text{W}$$

EXAMPLE 2 Calculate the power required to raise $0.15\,\text{m}^3$ of water per minute through a vertical distance of 35 m.

The mass of $1\,\text{m}^3$ of water is $10^3\,\text{kg}$.
The force required to raise this mass of water is

$$F = 0.15 \times 10^3 \times 9.81\,\text{N}$$

The power required is

$$P = \frac{F \times s}{t}$$

$$= \frac{0.15 \times 10^3 \times 9.81\,\text{N} \times 35\,\text{m}}{1 \times 60\,\text{s}}$$

$$= \frac{0.15 \times 10^3 \times 9.81 \times 35}{60}\,\frac{\text{N m}}{\text{s}} \quad \text{or} \quad \frac{\text{J}}{\text{s}}$$

$$= 858\,\text{W}$$

EFFICIENCY

The pump performing the operation of the last example has efficiency 72%. The power required to drive the pump is then

$$P = 858 \times \frac{100}{72} \quad \text{(multiply by 100 and divide by 72 to produce an increase)}$$

$$= 1192\,\text{W}$$

EXAMPLE 1 A d.c. motor has full load output of 5 kW. Under these conditions the input is 250 V, 26 A. Calculate the efficiency.

$$\text{Efficiency } \eta = \frac{\text{output power}}{\text{input power}}$$

$$= \frac{5 \times 1000}{250 \times 26} \quad (P = U \times I)$$

$$= 0.769 \text{ or } 76.9\% \text{ (after multiplying by 100)}$$

EXAMPLE 2 Calculate the current taken by a 10 kW, 460 V d.c. motor of efficiency 78%.

Output power $= 10\,\text{kW}$

Input power $= 10 \times \dfrac{100}{78}\,\text{kW} \quad$ (i.e. greater than output power)

$$= \frac{10 \times 100 \times 1000}{78}\,\text{W}$$

Since $P = U \times I$,

$$\frac{1000 \times 1000}{78} = 460 \times I$$

$$\therefore \qquad I = \frac{1\,000\,000}{460 \times 78}$$

$$= 27.9\,\text{A}$$

---

1. A force of 120 N is applied at right angles to the end of a bar 1.75 m long. Calculate the turning moment produced about a point at the other end of the bar.

2. Calculate the force required which when applied at right angles to the end of a bar 0.72 m long will produce a turning moment of 150 N m about a point at the other end.

3. Complete the following table, which refers to Fig. 36:

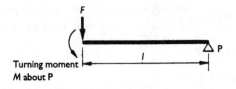

Fig. 36

F (newtons)	85		0.25	6.5	
l (metres)	0.35	1.2		0.125	2.75
M (newton metres)		50	0.15		500

4. A bar 1.5 m long is pivoted at its centre. A downward force of 90 N is applied at right angles 0.2 m from one end. Calculate the downward force to be applied at right angles to the bar at the opposite end to prevent it from rotating. Neglect the weight of the bar.

5. A bar 0.8 m long is pivoted at its centre. A downward force of 150 N is applied at right angles to the bar at one end. At what distance from the opposite end of the bar should a vertically downwards force of 200 N be applied to create equilibrium? Neglect the weight of the bar.

6. A force of 25 N is used to move an object through a distance of 1.5 m. Calculate the work done.

7. Energy amounting to 250 J is available to move an object requiring a force of 12.5 N. Through what distance will the object move?

8. Calculate the energy required to raise a load of 240 kg through a vertical distance of 8.5 m.

9. Calculate the energy required to raise 2.5 m^3 of water from a well 12.5 m deep.

10. A force of 0.15 N is used to move an object through 75 mm in 4.5 s. Calculate (a) the work done, (b) the power.

11. Calculate the power required to raise a load of 120 kg through a vertical distance of 5.5 m in 45 s.

12. Complete the following table, which refers to Fig. 31, page 105;

Distance between effort and pivot (m)	1	1.5	1.25		1.8
Distance between load and pivot (m)	0.125		0.15	0.10	0.20
Load (kgf)		200		390	225
Effort (kgf)	20		50	65	
Force ratio		5			

13. Complete the following table, which refers to Fig. 32, page 107:

Load to be raised (kgf)	250	320	420		500
Effort required (kgf)		150	75	80	
Vertical height (m)	3	4		2.4	1.8
Length of inclined plane (m)	6		5	5.4	4.2

14. A screwjack as illustrated in Fig. 33, page 108, has a thread of pitch 8 mm and a radius arm of length 0.5 m. Determine

(a) the effort required to raise a load of 1000 kgf,

(b) the load which an effort of 5.5 kgf will raise.

(c) What length of radius arm would be required to raise a load of 2500 kgf using an effort of 7.5 kgf?

15. Complete the following table, which refers to the wheel and axle illustrated in Fig. 34, page 109:

Radius of wheel R (cm)	25	16		17.5	30
Radius of axle r (cm)	8	6.5	6		8.5
Load G (kgf)	200		255	150	175
Effort E (kgf)		75	76.5	72.9	

16. A load of 275 kgf is to be raised using rope falls as illustrated in Fig. 35, page 111. Determine the effort required using each of the arrangements (b), (c), and (d). (Ignore friction.)

17. An effort of 85 kgf is applied to each of the arrangements in Fig. 35(b), (c), and (d), page 111. Ignoring friction, determine the load which could be raised in each case.

18. A motor and gear unit weighing 450 kgf is to be raised through a vertical distance of 2.5 m. It is proposed to use an inclined plane 4 metres long and a set of rope falls as in Fig. 35(d). Ignoring friction, determine the effort required.

19. A pump raises $0.15\,m^3$ of water per minute from a well 7.5 m deep. Calculate

(a) the power output of the pump;

(b) the power required to drive the pump, assuming an efficiency of 75%;

(c) the energy supplied to the pump in one hour.

20. A test on a d.c. motor produced the following results:

Input	240 V	Output	3200 W
	15 A		

Calculate the efficiency.

21. Calculate the full-load current of the d.c. motors to which the following particulars refer:

	Supply e.m.f. (V)	Output power (kW)	Efficiency (%)
(a)	240	1	68
(b)	480	15	82
(c)	200	2	74
(d)	250	4	75
(e)	220	10	78

22. A pump which raises $0.12\,m^3$ of water per minute through a vertical distance of 8.5 m is driven by a 240 V d.c. motor. Assuming that the efficiency of the pump is 72% and that of the motor is 78%, calculate the current taken by the motor.

23. A motor-generator set used for charging batteries delivers 24 A at 50 V. The motor operates from a 220 V supply and its efficiency is 70%. The generator is 68% efficient. Calculate the cost of running the set per hour at full load if the electrical energy costs 4.79p per unit.

24. A pumping set delivers $0.6\,m^3$ of water per minute from a well 5 m deep. The pump efficiency is 62%, that of the motor is 74%, and the terminal voltage is 234 V. Calculate

(a) the motor current;

(b) the cost of pumping $100\,m^3$ of water with energy at 5.18p per unit,

(c) the cross-sectional area of the copper cable which will supply the set from a point 50 m away with a voltage drop of not more than 6 V. (The resistivity of copper is $1.78 \times 10^{-8}\,\Omega\,m$.)

25. A d.c. motor at 460 V is required to drive a hoist. The load to be raised is 4000 kg at a speed of 0.2 m/s.

Calculate the minimum power of motor needed to do this work and also the current it would take, assuming the respective efficiencies of hoist gearing and motor to be 85% and 70%.

State the type of motor to be used, and give reasons for the choice.

26. A 50 m length of two-core cable of cross-section $70\,mm^2$ supplies a 240 V, 30 kW d.c. motor working at full load at 85% efficiency.

(a) Calculate the voltage drop in the cable.

(b) What steps would you take to reduce the voltage drop to half the above value, with the same load?

The resistivity of copper may be taken as $1.78 \times 10^{-8}\,\Omega\,m$.

27. A conveyor moves 400 kg upwards through a vertical distance of 14 m in 50 s. The efficiency of the gear is 38%. Calculate the power output of the driving motor. The motor is 78% efficient. Calculate the current it takes from a 250 V d.c. supply.

28. The bar in Fig. 37 is in equilibrium. The force F is
    (a) 4.8 N   (b) 2083 N   (c) 208.3 N   (d) 75 N

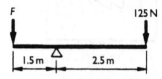

Fig. 37

29. A machine weighing 150 kgf is raised through a vertical distance of 1.5 m by forcing it up a sloping ramp 2.5 m long. Neglecting friction, the effort required is
    (a) 37.5 kgf   (b) 90 kgf   (c) 250 kgf   (d) 562.5 kgf

30. With reference to Fig. 34, page 109, if the radius of the large wheel is 25 cm and that of the axle is 8.5 cm, the load which could be raised by exerting an effort of 95 kgf is
    (a) 2794 kgf   (b) 279 kgf   (c) 32.3 kgf   (d) 323 kgf

Electrostatics

THE PARALLEL-PLATE CAPACITOR

An arrangement of two parallel metal plates each having cross-sectional area A (m^2) and separated by insulation of thickness d (m) is called a capacitor (Fig. 38). When it is connected to a d.c. supply of U volts it becomes charged, and the following facts are known:

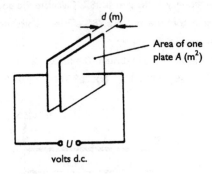

Fig. 38

(a) the quantity of charge in coulombs is

$$Q = CU \text{ where } C \text{ is the capacitance in farads (F)}$$

(b) the electric field strength is

$$E = U/d \text{ volts/metre}$$

(c) the energy stored is

$$W = \tfrac{1}{2}CU^2 \text{ joules}$$

(d) the capacitance of the arrangement is

$$C = \varepsilon_0 \varepsilon_r \frac{A}{d} \text{ farads}$$

where ε_0 is the permittivity of free space, with numerical value 8.85×10^{-12} (approximately $1/36\pi \times 10^{-9}$)

and ε_r is the relative permittivity of the insulator or dielectric

---

EXAMPLE Two parallel metal plates, each of area $0.01\,\text{m}^2$ and separated by a layer of mica 2 mm thick and of relative permittivity 6, are connected to a 100 V d.c. supply. Calculate

(a) the capacitance of the arrangement,
(b) the charge stored,
(c) the energy stored,
(d) the field strength in the dielectric.

(a) $$C = \varepsilon_0 \varepsilon_r \frac{A}{d}$$

$$= 8.85 \times 10^{-12} \times 6 \times 0.01 \times \frac{1}{2/1000}$$ (note conversion of 2 mm to m)

$$= 265.5 \times 10^{-12} \text{ farad}$$

$$= 265.5 \text{ picofarad (pF)}$$

$(1\,F = 10^6\,\mu F = 10^{12}\,pF)$

(b) $$Q = CU$$

$$= 265.5 \times 10^{-12} \times 100$$

$$= 26\,550 \times 10^{-12} \text{ coulombs}$$

$$= 0.026\,55 \times 10^{-6} \text{ coulombs}$$

$$= 0.026\,55 \text{ microcoulombs } (\mu C)$$

(c) $$W = \tfrac{1}{2}CU^2$$

$$= \tfrac{1}{2} \times 265.5 \times 100^2 \times 10^{-12}$$

$$= 132.75 \times 10^{-8} \text{ joules}$$

$$= 1.3275 \times 10^{-6} \text{ joules}$$

$$= 1.3275 \text{ microjoules } (\mu J)$$

(d) $$E = \frac{U}{d}$$

$$= \frac{100}{2/1000}$$

$$= 50\,000 \text{ volts/metre (U/m)}$$

CAPACITORS IN SERIES

If a number of capacitors of values C_1, C_2, C_3, etc. are connected in series (Fig. 39), they are equivalent to a single capacitor of value C given by

$$\frac{1}{C} = \frac{1}{C_1} + \frac{1}{C_2} + \frac{1}{C_3} + \ldots + \text{etc.}$$

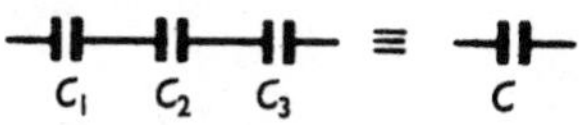

Fig. 39

When the arrangement is connected to a d.c. supply of V volts, the charge stored is the same on each capacitor and is equal to $Q = CU$.

EXAMPLE I Calculate the value of a capacitor which when connected in series with another of $20\,\mu F$ will give a resulting capacitance of $12\,\mu F$.

$$\frac{1}{C} = \frac{1}{C_1} + \frac{1}{C_2}$$

$$\frac{1}{12} = \frac{1}{20} + \frac{1}{C_2}$$

$$\therefore \quad \frac{1}{C_2} = \frac{1}{12} - \frac{1}{20}$$

$$\frac{1}{C_2} = \frac{5-3}{60}$$

$$= \frac{2}{60}$$

$$\therefore \quad C_2 = \frac{60}{2} = 30\,\mu F$$

The required value is thus $30\,\mu F$.

EXAMPLE 2 Capacitors of $4\,\mu F$, $6\,\mu F$, and $12\,\mu F$ are connected in series to a $300\,V$ d.c. supply. Calculate
(a) the equivalent single capacitor,
(b) the charge stored on each capacitor,
(c) the p.d. across each capacitor.

(a) $$\frac{1}{C} = \frac{1}{4} + \frac{1}{6} + \frac{1}{12}$$

$$= \frac{3+2+1}{12}$$

$$= \frac{6}{12}$$

$$C = \frac{12}{6} = 2\,\mu F$$

(b) $\left[\begin{array}{c}\text{Charged stored on each}\\\text{capacitor}\end{array}\right] = \left[\begin{array}{c}\text{charge stored on equivalent}\\\text{single capacitor}\end{array}\right]$

$$Q = CU$$
$$= 2 \times 300$$
$$= 600\,\mu C$$

(microcoulombs because capacitance is in microfarads)

(c) The p.d. on each capacitor is found by using the formula $Q = CU$, where C is the appropriate value of capacitance and Q is as calculated above.

Rearranging the formula gives

$$U = \frac{Q}{C}$$

Thus, for the $4\,\mu F$ capacitor,

$$U_4 = \frac{600}{4} = 150\,\text{V}$$

Similarly,

$$U_6 = \frac{600}{6} = 100\,\text{V}$$

and $\quad U_{12} = \frac{600}{12} = 50\,\text{V}$

(Note that these add up to 300 V.)

CAPACITORS IN PARALLEL

If a number of capacitors of values C_1, C_2, C_3 etc. are connected in parallel (Fig. 40), they are equivalent to a single capacitor of value C given by

$$C = C_1 + C_2 + C_3 + \ldots + \text{etc.}$$

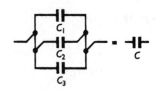

Fig. 40

When the arrangement is connected to a supply of V volts, the total charge is the sum of the charges stored separately on each capacitor; i.e., if Q is the total charge, then

$$Q = Q_1 + Q_2 + Q_3$$

where Q_1 is the charge on C_1 etc. and $Q_1 = C_1 V$ etc. and the voltage U is common to all the capacitors.

EXAMPLE   Capacitors of $4\,\mu F$ and $5\,\mu F$ are connected in parallel and charged to 20 V. Calculate the charge stored on each and the total stored energy.

The p.d. is the same for each capacitor. The charge on the $4\,\mu F$ capacitor is

$$Q_4 = C_4 U$$
$$= 4 \times 20$$
$$= 80\,\mu C$$

(microcoulombs because C is in microfarads)

Similarly,

$$Q_5 = C_5 U$$
$$= 5 \times 20$$
$$= 100\,\mu C$$

The total energy may be calculated either by finding the energy stored separately on each capacitor and adding, or by considering the total capacitance thus:

$$C = 4 + 5$$
$$= 9\,\mu F$$

and   total energy $W = \frac{1}{2} C U^2$
$$= \frac{1}{2} \times 9 \times 20^2$$
$$= 1800\,\mu J$$

(microjoules because C is in microfarads)

$$\varepsilon_O = 8.85 \times 10^{-12} \quad \left(\text{or} \ \frac{1}{36\pi} \times 10^{-9} \right)$$

1. Complete the following table, which refers to a certain variable capacitor:

Applied volts (U)	50		25	80	45
Capacitance (μF)		0.3	0.4		0.8
Charge (μC)	10	18		48	

2. A capacitor consists of two parallel metal plates each 0.1 m by 0.12 m and separated by a sheet of insulation having relative permittivity 7.5 and of thickness 1.5 mm. Calculate
   (a) its capacitance,
   (b) the charge and energy stored when the capacitor is charged to 75 V.

3. A parallel-plate capacitor consists of two parallel circular metal plates each 60 mm in diameter and separated by a layer of mica 0.75 mm thick. The relative permittivity of the mica is 6. Calculate (a) the capacitance of the arrangement, (b) the charge and energy stored when the p.d. between the plates is 100 V.

4. Calculate the diameter of circular plates required to produce a capacitance of 100 pF if the plates are separated by 0.5 mm of insulation of relative permittivity 5.

5. A certain capacitor has capacitance 12 pF with air between the plates. Calculate the capacitance when a sheet of insulating material of relative permittivity 7.5 is inserted between the plates. All other factors remain unchanged.

6. A parallel-plate capacitor is arranged so that the effective area of the plates can be varied from zero to 2.5×10^{-4} m^2. The distance between the plates is maintained constant at 1 mm. Plot a graph to show the variation in capacitance as the effective area is varied. (Take the relative permittivity of the dielectric as 1.)

7. A parallel-plate capacitor is arranged so that the separation of the plates in air can be varied between 0.5 mm and 2 mm. All the other factors remain constant. Plot a graph showing the variation in capacitance as the separation varies.

8. A capacitor of 100 pF is charged to a p.d. of 100 V. Calculate the charge and energy stored. Calculate also the energy stored if the distance

between the plates is (a) halved, (b) doubled. (Remember that the charge Q stored is constant; both C and V vary.)

9. Capacitors of 3 µF and 5 µF are connected in series to a 240 V d.c. supply. Calculate
   (a) the resultant capacitance,
   (b) the charge on each capacitor,
   (c) the p.d. on each capacitor,
   (d) the energy stored in each capacitor.

10. Calculate the value of a single capacitor equivalent to three 24 µF capacitors connected in series. What would be the value of ten 24 µF capacitors connected in series?

11. What value of capacitor connected in series with one of 20 µF will produce a resultant capacitance of 15 µF?

12. Three capacitors, of values 8 µF, 12 µF, and 16 µF respectively, are connected across a 240 V d.c. supply, (a) in series and (b) in parallel. For each case, calculate the resultant capacitance and also the potential difference across each capacitor.

13. Calculate the value of the single capacitor equivalent to the arrangement shown in Fig. 41.

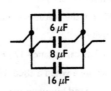

Fig. 41

14. A capacitor is formed from two parallel metal plates, each 0.15 m by 0.15 m, separated by a sheet of insulation material having relative permittivity 7.2 and of thickness 1.5 mm. The value of the capacitance is
    (a) 0.956 pF   (b) 956 F   (c) 956 pF   (d) 956 µF

15. A 12 µF capacitor is charged to 25 V. The energy stored is
    (a) 150 µJ   (b) 3750 µJ   (c) 3750 J   (d) 150 J

16. Capacitors of 2 µF and 4 µF are connected in series. When an additional capacitor is connected in series, the combined capacitance falls to 1 µF. The value of the third capacitor is
    (a) 4 µF   (b) 0.55 µF   (c) 0.25 µF   (d) 1.2 µF

17. Capacitors of 8 μF, 12 μF, and 20 μF are connected in parallel. For a total charge of 4000 μC to be stored, the voltage to be applied to the combination is

    (a) 0.01 V   **(b)** 15 480 V   **(c)** 100 V   **(d)** 1034 V

Answers

Exercise 1

1. 2768 W   **2.** 450 000 Ω   **3.** 37 mA   **4.** 3300 V

5. 596 Ω   **6.** 49.378 kW   **7.** 0.0165 A   **8.** 132 kV

9. 0.000 001 68 C   **10.** 0.724, W

11. (a) 0.000 000 36 Ω, (b) 1600 Ω, (c) 85 000 Ω, (d) 0.000 020 6 Ω,
     (e) 0.000 000 68 Ω

12. (a) 1850 W, (b) 0.0185 W, (c) 185 000 W, (d) 0.001 850 W, (e) 18.5 W

13. (a) 0.0674 V, (b) 11 000 V, (c) 240 V, (d) 0.009 25 V, (e) 6600 V

14. (a) 0.345 A, (b) 0.000 085 4 A, (c) 0.029 A, (d) 0.005 A, (e) 0.0064 A

15. 139.356 Ω   **16.** 5040 W

17. (a) 5.3 mA, (b) 18.952 kW, (c) 19.5 MΩ, (d) 6.25 µC, (e) 264 kV

18. 5.465 kW   **19.** 594 250 Ω   **20.** 0.0213 A   **21.** 0.000 032 5 C

22. 0.004 35 µF   **23.** (d)   **24.** (d)   **25.** (a)

26. (c)

Exercise 2

1. (a) 106 Ω, (b) 12.5 Ω, (c) 24 Ω, (d) 1.965 Ω, (e) 154.94 Ω, (f) 346.2 Ω, (g) 59.3 kΩ,
     (h) 2 290 000 Ω, (i) 0.0997 Ω, (j) 57 425 µΩ

2. (a) 22 Ω, (b) 2.35 Ω, (c) 1.75 Ω, (d) 2.71 Ω, (e) 1.66 Ω, (f) 13.42 Ω, (g) 6.53 Ω,
     (h) 1805 Ω, (i) 499 635 µΩ, (j) 0.061 MΩ

3. 3.36 Ω   **4.** 21.1 Ω   **5.** 9   **6.** 533 Ω, 19

7. 133.6 Ω, 30.4 Ω   **8.** 2.76 Ω   **9.** (c)   **10.** (d)

11. (b)   **12.** (b)

Exercise 3

1. (a) 1 Ω, (b) 1.58 Ω, (c) 3.94 Ω, (d) 1.89 Ω, (e) 2.26 Ω, (f) 11.7 Ω, (g) 6 Ω, (h) 5 Ω,
     (i) 10 Ω

2. (a) 16 Ω, (b) 6.67 Ω, (c) 7.2 Ω, (d) 6 Ω, (e) 42 Ω, (f) 2000 Ω, (g) 300 Ω, (h) 37.5 Ω,
     (i) 38 Ω, (j) 17.3 Ω

3. (a) 2.13 A, (b) 8.52 A   **4.** (a) 4.2 V, (b) 3.53 A, 26.4 A

5. 0.125 Ω   **6.** 25 A, 24.4 A, 21.6 A   **7.** 10.9 Ω

8. 40 A, 30 A   **9.** 20 A   **10.** 0.1 Ω

11. (a) 0.009 23 Ω, (b) 1.2 V, (c) 66.7 A, 50 A, 13.3 A

12. (a) $I_A = I_B = I_D = 2.5$ A, $I_C = 6.67$ A, (b) 9.17 A,
     (c) $U_A = 10$ V, $U_B = 12.5$ V, $U_C = 40$ V, $U_D = 17.5$ V, (d) 9.4 A

13. (d) **14.** (a) **15.** $0.365 \,\Omega$ **16.** 24 V, 12.4 V

17. (a) $0.0145 \,\Omega$, (b) 400 A **18.** 141 A, 109 A **19.** $0.137 \,\Omega$

20. $971 \,\Omega$ **21.** (a) 152 A, 121 A, 227 A, (b) 1.82 V, (c) 1.82 V

22. (b) **23.** (b)

Exercise 4

1. (a) $3.6 \,\Omega$, (b) $5 \,\Omega$, (c) 4 A in $9 \,\Omega$ resistance, 6 A in $6 \,\Omega$ resistance, 10 A in $1.4 \,\Omega$ resistance

2. (a) $2 \,\Omega$, (b) 12 V **3.** $2.25 \,\Omega$ **4.** $2.86 \,\Omega$

5. (a) $5.43 \,\Omega$, (b) 3.68 A **6.** 26.2 V

7. 0.703 A in $7 \,\Omega$ resistance, 0.547 A in $9 \,\Omega$ resistance

8. $11.7 \,\Omega$ **9.** 0.174 A, 0.145 A, 50.6 W **10.** $37.73 \,\Omega$

11. $25.4 \,\Omega$ **12.** (a) 20.31 A, (b) $0.23 \,\Omega$, (c) 2.64 kW

13. 216.3 V, 213.37 V

14. (a) $5 \,\Omega$, (b) $I_A = 3$ A, $I_B = 12$ A, $I_C = 15$ A **15.** $6 \,\Omega$

16. (a) 20 A, (b) $2 \,\Omega$, (c) 56 A **17.** 2.96 V **18.** 4.25 V

19. 6.96 V **20.** $0.0306 \,\Omega$

21. (a) 1.09 A, 0.78 A, (b) 5.44 V, (c) 1.05 W

22. (a) 4.37 V, (b) 0.955 W, 0.637 W

23. 88.7 V **24.** 107 V **25.** 117 V, 136 W, 272 W

26. $2.59 \,\Omega$ **27.** (a) $4.4 \,\Omega$, (b) 15.9 V, (c) $1.72 \,\Omega$

28. (b) **29.** (d) **30.** (a)

Exercise 5

1.

U (volts)	10	20	30	40	50
I (amperes)	1	2	3	4	5
R (ohms)	10	10	10	10	10

2. 36 V

3.

U (volts)	240	240	240	240	240
I (amperes)	12	6	4	3	2.4
R (ohms)	20	40	60	80	100

4. 3 A, $23 \,\Omega$

5.

U (volts)	100	100	96	56	96	132	84	144	121	63
I (amperes)	10	10	12	7	8	12	7	12	11	9
R (ohms)	10	10	8	8	12	11	12	12	11	7

6.

I (amperes)	100	10	10	0.1	0.1	0.1	100	0.001	0.1	200
R (ohms)	0.1	1000	0.1	1000	0.1	1000	0.1	2000	2000	0.01
U (volts)	10	100	1	10	0.01	100	10	20	200	2

7.

R (ohms)	480	14	500	16	110	0.07	12	500	0.75	15
I (amperes)	0.5	15	0.05	6	1.2	0.9	0.7	0.2	8	8
U (volts)	240	210	25	96	132	0.063	8.4	100	6	120

8. 2.41 V

9. (a) 0.33 Ω, (b) 0.17 Ω, (c) 0.12 Ω, (d) 0.7 Ω, (e) 0.045 Ω

10. 11 A

11. Section SA 2.916 V, section AB 4.253 V, section AC 2.322 V

volts at A = 47.08 V, volts at B = 42.83 V, volts at C = 40.51 V

12.

Rated current (A)	5	15	30	60	100
Minimum fusing current (A)	7	21	42	84	140

13.

Rated current (A)	5	15	30	60	100
Minimum fusing current (A)	6	18	36	72	120

14. (a) 0 (fuse will blow), (b) 216.48 V, (c) 15.33 V, (d) 184 V, (e) 184 V, (f) 191.6 V

15. (b)   **16.** (c)   **17.** (c)   **18.** (a)

Exercise 6

1. (a) 0.533 A, (b) 0.632 A, (c) 0.563 A, (d) 0.612 A, (e) 0.593 A

2. 63.2 mA, 92.3 mA, 120 mA, 171 mA, 300 mA,

3. (a) 19.2 A, (b) 9.6 A   **4.** 51.4 Ω

5. (a) 27.7 V, (b) 17.2 V, (c) 17.5 V, (d) 14.8 V, (e) 23.1 V

6.

U (volts)	20	40	101	80	120	60
I (amperes)	0.2	0.2	0.2	0.2	0.195	0.2
R (ohms)	100	200	505	400	615	300

(a) 30 V, (b) 64 V

Exercise 7

1. 3.96 V   **2.** 36 A   **3.** 59.1 A   **4.** 80.3 A

5. (a) 2.76 mV/A/m, (b) 227.24 V

6. (a) 40 A, (b) 3.97 mV/A/m, (c) 16 mm^2, (d) 3.53 V

7. (a) 40 A, (b) 41.2 A, (c) 10 mm^2, (d) 4.4 mV/A/m, (e) 3.92 V

Exercise 8

1.

P (watts)	1440	3000	1600	1000	20	1000	2350	1080
I (amperes)	6	12	6.67	150	0.2	5.45	5.1	4.5
U (volts)	240	250	240	6.67	100	220	460	240

2. 30 W   **3.** 4370 W

4. (a) 13.04 A, (b) 6.53 A, (c) 1.96 A, (d) 15.22 A, (e) 30.44 A, (f) 0.26 A, (g) 0.43 A, (h) 8.7 A, (i) 3.26 A, (j) 0.065 A

5. 108 V   **6.** 203 W   **7.** (a) 18.75 A, (b) 20.45 A   **8.** 748 W

9. 6 A   **10.** 13.04 A   **11.** 633 W   **12.** 0.2 A, 23 W

13. 17.4 A   **14.** (a) 6.413 kW, (b) 83.7 W, (c) 222 V

15. 29.2 Ω   **16.** (a) 460 V, (b) 7.05 kW, (c) 76.9 W

17. (a) 62.1 W, 70.9 W, (b) 286 W, 250 W

18. 453 W, 315 W   **19.** 40.5 V, 0.78 Ω, 750 W

20. (a) 434 V, (b) 15.6 kW, (c) 216 W, (d) 433 V
21. (c)   **22.** (b)   **23.** (c)   **24.** (a)

Exercise 9

1.

Power (W)	1500	200	1800	1440	1000	2640	100	42.25
Current (A)	10	5	15	12	4.2	11.49	0.42	1.3
Resistance (Ω)	15	8	8	10	56.7	20	567	25

2. 130 W **3.** 29.4 W **4.** 5 A **5.** 1.73 kW
6. 576 Ω **7.** (a) 6.21 V, (b) 71.4 W **8.** 530 W
9. 20 A **10.** 59.4 W **11.** 125 Ω **12.** 63.4 W
13. 419 W **14.** 0.248 W **15.** 0.8 A

Exercise 10

1. 72 W **2.** 130 Ω **3.** 4 Ω **4.** 120 V

5.

Power (W)	128	100	60	125	768	1800	42.24	36
Voltage (V)	80	240	250	50	240	220	3.5	12
Resistance (Ω)	50	576	1042	20	75	26.9	0.29	4

6. 15 V **7.** 557 W **8.** 52.9 Ω **9.** 170 W **10.** 161 Ω
11. (a) 28.8 Ω, (b) 19.2 Ω, (c) 16.5 Ω, (d) 128 Ω, (e) 960 Ω, (f) 8.23 Ω, (g) 576 Ω, (h) 38.4 Ω, (i) 76.8 Ω, (j) 14.4 Ω,
12. 79.1 V **13.** 4.19 A **14.** (a) 0.149 A, (b) 29.8 W
15. 115 Ω, 28.8 Ω, 500 W, 2000 W **16.** 19 A, 126 W, 81 A, 538 W
17. 750 W, 3000 W

18.

Power (W)	25.6	250	360	400	600	960
Current (A)	0.8	2.5	3	3.15	3.87	4.9
Resistance (Ω)	40	40	40	40.3	40	40

(a) 550 W approx., (b) 4.4 A approx.

19.

Power (W)	3000	2000	1661	750	420	180
Voltage (V)	240	200	180	120	89.6	58.8
Resistance (Ω)	19.2	20	19.5	19.2	19.1	19.2

(a) 175 V, (b) 3200 W approx.

22. (d) **23.** (b) **24.** (d) **25.** (c)

Exercise 11

1. 61.3 C **2.** 200 s

3.

Current (A)	0.025	15.3	0.55	11.7	8.5
Time (s)	125	7.5	1910	360	63.5
Charge (C)	3.13	115	1050	4200	540

4. (a) 54 000 C, (b) 15 Ah **5.** 31.3 **6.** 14 400 J
7. 12 150 J **8.** (a) 129 600 C, 36 Ah, (b) 31 104 000 J, 8.64 kWh
9. £6.44 **10.** £4.14 **11.** 33.97p

12. (a) 23.94p, (b) 1.2p, (c) 3.6p, (d) 2.79p, (e) 0.56p

13. (a) 49.03 h, (b) 1.48 h, (c) 14.7 h, (d) 10.9 h, (e) 117.64 h

14. £232.03   **15.** £62.71   **16.** £162.84   **17.** £62.94

18. £6.10   **19.** 54p   **20.** £1.70   **21.** 28.8p

22. 81.76p   **23.** £1.72   **24.** £89.81

25. Electricity £26.44, coal £31.70   **26.** £378.65   **27.** (a)

28. (c)   **29.** (c)   **30.** (b)

Exercise 12

1. (a) 10 V, (b) 16.6 V, (c) 4.4 V   **2.** 350 W

3. (a) 224.25 V, (b) 235.75 V   **4.** 235 V, 225 V   **5.** 3447 W

6. 3%   **7.** 0.69%   **8.** (a) 3640 W, (b) 109 W, (c) 2.9%

9. 0.276 Ω **10.** 242.4 V, 237.6 V   **11.** 3.2 A   **12.** 74.3%

13. 69.4%   **14.** 72.5%   **15.** 4.37 kW   **16.** 14.5 A

17. (a) 8.94 W, (b) 13.9%   **18.** (a) 5%, (b) 2708 W, (c) 9.73%

19. (a) 1.53%, (b) 0.74%   **20.** 38   **21.** (c)

22. (a)

Exercise 13

1. 10.5 m^2 **2.** 0.14 m

3.

Length (m)	6	3	12	8	12
Breadth (m)	2	2	7	4	4
Perimeter (m)	16	10	38	24	32
Area (m^2)	12	6	84	32	48

4. 10.5 m^2 **5.** 19 m^2

6.

Base (m)	0.5	4	1.5	11.25	0.3
Height (m)	0.25	4.5	2.2	3.2	0.12
Area (m^2)	0.625	9	1.65	18	0.018

7.

Area (m^2)	0.015	0.25×10^{-3}	7.5×10^{-3}	0.000 29	0.0016
Area (mm^2)	15×10^3	250	7500	290	1.6×10^3

8.

Diameter	0.5 m	0.318 m	0.7927 m	2.76 mm	4 mm
Circumference	1.571 m	1.0 m	0.252 m	8.67 mm	12.57 mm
Area	0.196 m^2	0.079 m^2	0.5 m^2	5.98 mm^2	12.57 mm^2

9. 0.331 m^2 (575 mm $\times$ 575 mm), 9 rivets

10. 0.633 m^2 (660 mm $\times$ 958 mm), 2.64 m angle, 80 rivets   **11.** 19.6 m

12.

No. and diameter (mm) of wires	1/1.13	1/1.78	7/0.85	7/1.35	7/2.14
Nominal c.s.a. (mm^2)	1	2.5	4	10	25

13.

Nominal overall diameter of cable (mm)	2.9	3.8	6.2	7.3	12.0
Nominal overall cross-sectional area (mm)	6.6	11.3	30.2	41.9	113

14. (a) 133 mm², (b) 380 mm², (c) 660 mm²

15.

	Permitted number of pvc cables in trunking of size (mm)		
Cable size	50 × 37.5	75 × 50	75 × 75
16 mm²	20	40	60
25 mm²	13	27	40
50 mm²	8	15	22

16. 75 mm × 50 mm or 100 mm × 37.5 mm

17. 8036 mm² (about 90 mm × 90 mm) use 100 mm × 100 mm

18. (c)   **19.** (b)   **20.** 75 mm × 75 mm trunking

21. 19 pairs can be added

22. 25 mm conduit, draw in box after second bend

23. (a) 32 mm conduit, (b) adequate room exists but re-calculation of new and existing cable ratings will be necessary

24. (a) 50 mm × 50 mm or 75 mm × 38 mm trunking, (b) 32 mm conduit, (c) difficulty may result when extending from stop-end of 75 mm × 38 mm trunking

Exercise 14

1. 45.5 mm³   **2.** 0.147 mm³

3. (a) 0.375 mm³, 3.25 m², (b) 0.0628 mm³, 0.88 m²

4. 1125 × 10⁻⁶ m³, 9.71 kg   **5.** 491 × 10⁻⁶ m, 4.24 kg

6. 2438 m³, 3120 kg   **7.** 96.6 m   **8.** 1.075 kg   **9.** 297 litres, 297 kg

10. 292 litres, 248 kg

Exercise 15

1. 1.2 Ω   **2.** 0.222 Ω   **3.** 6 mm²   **4.** 10 mm²

5. 43.2 m   **6.** 4.17 V, 6.66 V   **7.** 0.001 48 Ω   **8.** 6 mm

9. 0.623 Ω   **10.** 1.71 V   **11.** 48 m   **12.** 0.071 Ω

13. 47.4 m   **14.** 0.006 Ω   **15.** 25 mm   **16.** 2.5 mm²

17. (a) 1.187 Ω, (b) 0.297 Ω, (c) 0.178 Ω, (d) 0.051 Ω, (e) 0.0356 Ω

18. (a) 0.155 Ω, (b) 0.111 Ω, (c) 0.34 Ω

19. (a) 1.32 Ω, (b) 0.528 Ω, (c) 8.8 Ω   **20.** 0.15 Ω   **21.** 2.63 Ω

22. (a)   **23.** (c)   **24.** (a)

Exercise 16

1. (a) £12.20, (b) £207.40   **2.** (a) £1.32, (b) £322.95

3. (a) £2.83, (b) £467   **4.** £100.11   **5.** £84.60

6. £268.08   **7.** £535.04   **8.** (a) £1939.87, (b) £2230.85

9. £188.25   **10.** £672.25   **11.** (a) £297.90, (b) £349.98

12. (a) (i) £427.73, (ii) 74.85, (b) (i) £531.72, (ii) £93.05,
(c) (i) £1221.66, (ii) 213.70, (d) (i) £62.87, (ii) £11.00

13. (a) £558.51, (b) £625.81, (c) £762.07, (d) £872.57, (e) £2787.85

14. (c)   **15.** (b)

Exercise 17

1. 29.3×10^6 J   **2.** 47.7×10^6 J   **3.** 1 MJ   **4.** 41.9 °C

5. 57 °C   **6.** 62.3×10^6 J   **7.** 11.17 kW h   **8.** 51.4 °C

9. (a) 198 min, (b) 242 min   **10.** 5169 W   **11.** 32.8p, 105 min

12. 71.4%   **13.** (a) 61 min, (b) 66.7 min   **14.** 228 min

15. 13.43 kW, (a) £1.39, (b) 4.29 Ω   **16.** 373 min   **17.** 64.2 °C

18. (a) 113 A, (b) 2.12 Ω, (c) £1.49   **19.** 78.9%   **20.** 81 min

21. (c)   **22.** (d)

Exercise 18

1. (a) 1.25 mWb, (b) 0.795

2.

Wb	0.025	0.035	0.059	0.74	0.85
mWb	25	35	59.5	740	850
μWb	25 000	35 000	59 500	740 000	850 000

3. 0.0688 Wb   **4.** 0.792 T   **5.** 0.943 mWb   **6.** 1.919 N

7. 0.778 T

8.

Flux density (T)	0.95	0.296	1.2	0.56	0.706
Conductor length (m)	0.035	0.12	0.3	0.071	0.5
Current (A)	1.5	4.5	33.3	0.5	85
Force (N)	0.05	0.16	12	0.02	30

13. 0.08 V   **14.** 2.56 m/s   **15.** 180 V   **16.** 102 V

17. 6.5 V   **18.** 0.15 V   **19.** 100 H   **20.** 1.75 H

21. 50 V   **22.** 66.7 A/s   **23.** 30 H   **24.** 80 V

25. (a)   **26.** (c)   **27.** (b)   **28.** (c)

29. (a)   **30.** (c)   **31.** (d)

Exercise 19

1. 240 V   **2.** 13.1 A, 14.1 A   **4.** 25 mA, 28.9 mA

5. (a) 0.347 A, 0.429 A, (b) 18.4 W, (c) 2209 J

7.

Voltage (V)	240	238	242	110
Current (A)	3	5	10.2	4.8
Power (W)	576	893	1382	338
Power factor	0.8	0.75	0.56	0.64

8. (c)   **9.** (b)   **10.** (a)

Exercise 20

1. 17.4 A
2. (a) 870 T, (b) 6.52 V
3. 66.7 V
4. 236 V, 308 A
5. 1900, 144
6. (c)
7. (d)
8. (c)

Exercise 21

1. 210 N m
2. 208 N

3.
F (N)	85	41.7	0.25	6.5	182
l (m)	0.35	1.2	0.6	0.125	2.75
M (N m)	29.8	50	0.15	0.813	500

4. 66 N
5. 0.3 m
6. 37.5 J
7. 20 m
8. 20 000 J
9. 306 400 J
10. (a) 11.5 mJ, (b) 2.5 mW
11. 144 W

12.
Distance between effort and pivot (m)	1	1.5	1.25	0.6	1.8
Distance between load and pivot (m)	0.125	0.3	0.15	0.1	0.2
Load (kgf)	160	200	416.5	390	225
Effort (kgf)	20	40	50	65	25
Force ratio	8	5	8.33	6	9

13.
Load to be raised (kgf)	250	320	420	180	500
Effort required (kgf)	125	150	75	80	214.3
Vertical height (m)	3	4	0.89	2.4	1.8
Length of inclined plane (m)	6	8.53	5	5.4	4.2

14. (a) 2.55 kgf, (b) 2160 kgf, (c) 0.424 m

15.
Radius of wheel R (cm)	25	16	20	17.5	30
Radius of axle r (cm)	8	6.5	6	8.5	8.5
Load (kgf)	200	185	255	150	175
Effort (kgf)	64	75	76.5	72.9	49.6

16. (b) 137.5 kgf, (c) 91.7 kgf, (d) 68.8 kgf
17. (b) 170 kgf, (c) 255 kgf, (d) 340 kgf
18. 70.3 kgf
19. (a) 184 W, (b) 245 W, (c) 882×10^3 J
20. 88.9%
21. (a) 6.1 A, (b) 38.1 A, (c) 13.5 A, (d) 21.3 A, (e) 58.3 A
22. 1.19 A
23. 12.58p
24. (a) 4.57 A, (b) 15.35p, (c) 1.36 mm^2 (1.5 mm^2)
25. 13.2 kW (15 kW), 28.7 A
26. (a) 3.74 V
27. 2.892 kW, 14.8 A
28. (c)
29. (b)
30. (c)

Exercise 22

1.
Applied volts	50	60	25	80	45
Capacitance (μF)	0.2	0.3	0.4	0.6	0.8
Charge (μC)	10	18	10	48	36

2. (a) 531 pF, (b) 0.0398 µC, 1.49 µJ **3.** (a) 200 pF, (b) 0.02 µC
4. 37.9 mm **5.** 90 pF **8.** 0.01 C, 0.5 µJ, (a) 0.25 µJ, (b) 1 µJ
9. (a) 1.88 µF, (b) 450 µC, (c) 150 V, 90 V, (d) 0.34 J, 0.02 J
10. 8 µF, 2.4 µF **11.** 60 µF
12. (a) 3.7 µF, 111 V, 74 V, 56 V, (b) 36 µF, 240 V **13.** 30 µF
14. (c) **15.** (b) **16.** (a) **17.** (c)